SpringerBriefs in Electrical and Computer Engineering

SpringerBriefs present concise summaries of cutting-edge research and practical applications across a wide spectrum of fields. Featuring compact volumes of 50 to 125 pages, the series covers a range of content from professional to academic. Typical topics might include: timely report of state-of-the art analytical techniques, a bridge between new research results, as published in journal articles, and a contextual literature review, a snapshot of a hot or emerging topic, an in-depth case study or clinical example and a presentation of core concepts that students must understand in order to make independent contributions.

More information about this series at http://www.springer.com/series/10059

Cam Nguyen · Youngman Um

Multiband Dual-Function CMOS RFIC Filter-Switches

 Springer

Cam Nguyen
Texas A&M University
College Station, TX, USA

Youngman Um
Peregrine Semiconductor
Austin, TX, USA

ISSN 2191-8112   ISSN 2191-8120   (electronic)
SpringerBriefs in Electrical and Computer Engineering
ISBN 978-3-030-46247-5   ISBN 978-3-030-46248-2   (eBook)
https://doi.org/10.1007/978-3-030-46248-2

This Springer imprint is published by the registered company Springer Nature Switzerland AG
The registered company address is: Gewerbestrasse 11, 6330 Cham, Switzerland

Contents

About the Author

Dr. Nguyen joined the Department of Electrical and Computer Engineering, Texas A&M University in December 1990, after working for more than twelve years in industry, and is currently the Texas Instruments Endowed Professor. He was Program Director at the National Science Foundation during 2003–2004, where he was responsible for research programs in RF electronics and wireless technologies. From 1979 to 1990, he held various engineering positions in industry, serving as a Microwave Engineer with the ITT Gilfillan Company, a Member of Technical Staff with Hughes Aircraft (now Raytheon), a Technical Specialist with Aeroject ElectroSystems, a Member of Professional Staff with Martin Marietta (now Lockheed Martin), and a Senior Staff Engineer and Program Manager with TRW (now Northrop Grumman). While in industry, he led numerous microwave and millimeter-wave activities and developed many microwave and millimeter-wave hybrid and monolithic integrated circuits and systems up to 220 GHz for communications, radar, and remote sensing. His research group at Texas A&M University is currently focusing on CMOS/BiCMOS RFICs and systems, microwave and millimeter-wave ICs and systems, and ultra-wideband devices and systems for wireless communications, radar, and sensing—developing not only individual components, but also complete systems including design, signal processing, integration, and testing. He has published more than 315 papers, written seven books and six book chapters, and given more than 200 conference presentations and numerous invited presentations. Professor Nguyen was the Founding Editor-in-Chief of Sensing and Imaging: An International Journal published by Springer and the Founding Chairman of the International Conference on Subsurface Sensing Technologies and Applications.

Chapter 1
Introduction

Switches are important components found in many RF systems operating at microwave and millimeter-wave frequencies—from single-pole single-throw (SPST) to multi-pole multi-throw switches with single-pole double-throw (SPDT) and transmit/receive (T/R) switches being among the most popular ones. They can be used as a stand-alone component or integrated within subsystems or systems—for instance, T/R switches used in systems to accommodate the operation of transmitters and receivers sharing a common antenna. Switches are also used as an integral part of a component to enable the component to perform a particular function such as a SPST switch in a RF pulse-former component to modulate a continuous-wave (CW) signal to form a RF pulse signal. More complex multi-pole multi-throw switches can be employed to achieve various system functions like transmission and reception of dual-polarized signals using a single dual-polarized antenna. Besides the fundamental function of switching seen in commonly used switches, other functions such as filtering and attenuation can also be incorporated into a switch to form a multi-function switch like a filter-switch having concurrent switching and filtering functions or a switching-attenuator possessing both attenuation and switching at the same time.

Typical RF switches are designed only for switching purposes without considering frequency selectivity or filtering. A separate filter is normally used with a switch to provide a filtering function. For instance, a band-pass filter is typically needed between the antenna and transmit/receive (T/R) switch to suppress unwanted signals arriving from the antenna before going to the receiver's low-noise amplifier (for receiving) and coming out of the transmitter's power amplifier before reaching the antenna (for transmitting). This conventional design approach leads to higher insertion loss, larger size and higher cost for RF systems. To overcome these problems, it is logical to combine filter and switching functions in the design to produce a single circuit that is capable of simultaneous switching and filtering, while keeping the size comparable to that of a corresponding switch without filtering capability.

© The Author(s) 2020   1
C. Nguyen and Y. Um, *Multiband Dual-Function CMOS RFIC Filter-Switches*,
SpringerBriefs in Electrical and Computer Engineering,
https://doi.org/10.1007/978-3-030-46248-2_1

RF systems operating over multiple bands provide numerous advantages and have more capabilities as compared to their single-band counterparts for communications and sensing. The ability of operating multiple bands increases the diversity of RF systems for sensing and communication functions at multiple frequencies. Moreover, achieving concurrent functions over multiband enables a single RF system to be used at multiband simultaneously—avoiding the need of physically combining separate RF systems, each working in an individual band, together, which is difficult (and expensive) to realize in practice—particularly when many bands are involved. This leads to optimum size, cost and power consumption, and ease in realization for the system, besides possibly enhanced performance, and hence is beneficial for RF applications. True concurrent multiband RF systems require many of their constituent components to work concurrently in multiple bands. Concurrent multiband switches are some of them and their designs are thus crucial for successful developments of multiband RF systems.

Microwave and millimeter-wave integrated circuits can be implemented using silicon-based CMOS (or related BiCMOS) radio-frequency integrated circuits (RFICs) and microwave monolithic integrated circuits (MMICs) based on III-V compound semiconductor devices. CMOS RFICs have lower cost and better abilities for direct integration with digital ICs (and hence better potential for complete system-on-a-chip) as compared to those using III-V compound semiconductor devices. CMOS RFICs are also small and have low power consumption, making them suitable for portable battery-operated systems. CMOS RFICs are thus very attractive for RF systems and, in fact, the principal choice for commercial wireless markets. With respect to switches, Gallium Arsenide (GaAs) semiconductor transistors have been employed extensively taking advantages of their low on-state resistance and off-state capacitance, which are important for low insertion loss and high isolation in switching, and high linearity, needed for large-signal application, at high frequencies. Silicon-On-Insulator (SOI) technology has also been used for switches. It is implemented on high resistivity substrate with buried oxide layers serving to reduce capacitive coupling with substrate, leading to SOI switches with low insertion loss and high isolation [1–4]. Silicon-based CMOS technology has advanced significantly in the past two decades, resulting in improved processes leading to CMOS switches with better performance. Improved performance, coupled with lower cost, smaller power consumption, and most importantly ease in realizing single-chip systems, have make CMOS switches not only an alternate, but also important, switching solution for RF systems.

This book presents the theory, analysis and design of dual-band band dual-function CMOS RFIC filter-switches capable of simultaneous switching and filtering, which are relevant for advanced multiband RF systems. The book contains 6 chapters. The first chapter gives the introduction and background of switches. The second chapter covers the fundamentals of band-pass, high-pass and low-pass filters as well as dual-band band-pass filters, which the dual-band dual-function filter-switches addressed in this book are based upon. The third chapter presents the fundamentals of switches and various switch architectures including SPST, SPDT, and T/R switches. The fourth chapter discusses the fundamentals and models of MOSFETs used in the design of switches along with the deep n-well technique for improved switch performance. The fifth chapter presents the core of this book, which is the designs, simulations and measurements of various CMOS dual-band dual-function SPDT and T/R switches capable of concurrent switching and filtering, as examples to illustrate the design of multiband multi-function filter-switches. These dual-function filter-switches operate in two different frequency bands centered at around 40 and 60 GHz and 24 and 60 GHz. It is noted that 24, 40 and 60 GHz were chosen for our specific system and applications and herein serve as representative frequencies for demonstrations of our designs in dual-band dual-function filter-switches capable of simultaneous switching and filtering. Other frequencies can of course be used to design multi-band multi-function switches and other components for different systems and applications. Finally, summary and conclusion are given in Chap. 6.

References

1. Im D, Kim B-K, Im D-K, Lee K (2015) A stacked-FET linear SOI CMOS cellular antenna switch with an extremely low-power biasing strategy. IEEE Trans Microw Theor Tech 63(6):1964–1977
2. Wang XS, Yue CP (2014) A dual-band SP6T T/R switch in SOI CMOS With 37-dBm for GSM/W-CDMA handsets. IEEE Trans Microw Theor Tech 62(4):861–870
3. Tinella C, Fournier JM, Belot D, Knopik V (2003) A high-performance CMOS-SOI antenna switch for the 2.5–5-GHz band. IEEE J Solid-State Circ 38(7):1279–1283
4. Tombak A, Carroll MS, Kerr DC, Pierres J-B, Spears E (2013) Design of high-order switches for multimode applications on a silicon-on-insulator technology. IEEE Trans Microw Theor Tech 61(10).3639–3649

Chapter 2
Multi-band Band-Pass Filters

2.1 Introduction

A switch can be designed to have low-pass, high-pass, band-pass, or band-stop filtering characteristics, making it to function as a low-pass, high-pass, band-pass, or band-stop filter-switch, respectively. Such filter-switches can be designed by implementing corresponding filter topology for the switch's schematic. Filters therefore form the foundation for filter-switches. There are four types of filters: low-pass filter, high-pass filter, band-stop filter, and band-pass filter, from which switches with corresponding filtering function can be designed.

Multi-band band-pass filters allow filters to work concurrently in multiple pass-bands. Achieving concurrent filtering functions over multiband enables one single band-pass filter to be used at multi-band simultaneously—avoiding the need of physically combining separate band-pass filters, each working at an individual pass-band, together, which is difficult (and expensive) to realize in practice—particularly when many bands are involved and many filters are needed. Effectively, multi-band band-pass filters integrate multi-band together electrically besides physically. They are different from wide-band band-pass filters covering multiple pass-bands in a single bandwidth covering both desired and undesired pass-bands in the multi-band, in which they consist of multiple individual passbands separated from each other by stop-bands. In other words, multi-band band-pass filters only pass signals within their desired multiple pass-bands and reject others in between. Although, only dual-pass-band band-pass filters are covered in this chapter and only dual-pass-band filter-switches are presented in this book, other multi-band filters and filter-switches having different filtering function (e.g., band-stop) and/or more than two bands can also be designed following similar approaches.

The designs of the dual-band band-pass filter-switches presented in this book are based on multi-band band-pass filters, which are evolved from (single-band) band-pass filters. The theories of band-pass filters and multi-band band-pass filters are essential for the design and understanding of multi-band band-pass filter-switches.

© The Author(s) 2020
C. Nguyen and Y. Um, *Multiband Dual-Function CMOS RFIC Filter-Switches*,
SpringerBriefs in Electrical and Computer Engineering,
https://doi.org/10.1007/978-3-030-46248-2_2

This chapter presents the fundamentals of low-pass, high-pass, band-pass, and band-stop filters, extracted from [1, 2], which the filter-switches presented in this book are based upon. It also addresses the dual-band band-pass filters that the dual-band band-pass filter-switches presented in Chap. 5 are based on.

2.2 Low-Pass Filter

2.2.1 Prototypes

Low-pass filter is the most fundamental one and its prototypes, as shown in Fig. 2.1, form the basis for all of the filters (including low-pass filters), from which these filters are derived. The low-pass filter prototypes shown in Fig. 2.1 are dual of each other and produce identical responses. In these prototypes, either end can be used as the source or load impedance due the reciprocity of the passive network. The prototype elements are designated with common variables (g's) as indicated in Fig. 2.1 and defined as following:

g_k $(k = 1, 2, ..., n)$ = inductance of a series inductor or capacitance of a shunt capacitor

$$g_o = \begin{cases} source\ resistance\ R_o\ if\ g_1 = C_1 \\ source\ conductance\ G_o\ if\ g_1 = L_1 \end{cases}$$

$$g_{n+1} = \begin{cases} load\ resistance\ R_{n+1}\ if\ g_n = C_n \\ load\ conductance\ G_{n+1}\ if\ g_n = L_n \end{cases}$$

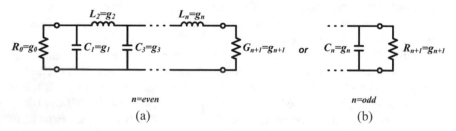

Fig. 2.1 Low-pass filter prototypes with even (**a**) and odd (**b**) number of elements n. Reprinted, with permission, from [2]

2.2.2 *Prototype Elements*

The most fundamental low-pass filter responses, which describe attenuation as a function of frequency, are maximally flat and Chebyshev responses. The prototype elements described in the following two sections are based on the maximally flat and Chebyshev responses.

2.2.2.1 Maximally Flat Prototype Elements

We consider the source and load terminations as resistors and the insertion loss (or maximum pass-band loss) for maximally-flat low-pass filter is 3 dB as typically the case. We also assume that all values are normalized with respect to the source termination and the cutoff frequency ω'_c; that is, $g_o = 1$ and $\omega'_c = 1$ rad/s. The elements of the maximally flat low-pass filter prototypes can be determined from the following equations:

$$g_o = 1$$
$$g_k = 2\sin\left[\frac{(2k-1)\pi}{2n}\right], \quad k = 1, 2, \dots, n$$
$$g_{n+1} = 1 \tag{2.1}$$

where g_k is in Henry (H) for inductance and Farad (F) for capacitance. It is noted that both impedance and frequency normalization are used in (2.1), which is necessary for generalization of the results. The elements for actual filters can be easily obtained by de-normalizing or scaling as discussed later.

2.2.2.2 Chebyshev Prototype Elements

As for the maximally-flat filter prototypes, it is assumed that the source and load terminations are resistors, and $g_o = 1$ and $\omega'_c = 1$ rad/s. The elements of the Chebyshev low-pass filter prototypes can be determined from the following equations:

$$g_o = 1$$
$$g_1 = \frac{2a_1}{p}$$
$$g_k = \frac{4a_{k-1}a_k}{a_{k-1}a_k}, \quad k = 1, 2, \dots, n$$
$$g_{n+1} = \begin{cases} 1, & n\ odd \\ \coth^2\left(\frac{q}{4}\right), & n\ even \end{cases} \tag{2.2}$$

where

$$q = \ln\left(\coth \frac{\alpha_{dB}}{17.37}\right)$$

$$p = \sinh \frac{q}{2n}$$

$$a_k = \sin\left[\frac{(2k-1)\pi}{2n}\right]$$

$$b_k = p^2 + \sin^2\left(\frac{k\pi}{n}\right)$$

with α_{dB} being the pass-band ripple or insertion loss in dB, g_k is in Henry (H) for inductance and Farad (F) for capacitance. The values calculated from (2.2) are normalized with respect to the source termination ($g_o = 1$) and cutoff frequency ($\omega'_c = 1$ rad/s).

It is noted that for the maximally flat prototypes, the source and load impedances are equal regardless whether the filter order n is odd or even while, for the Chebyshev prototypes, the source and load impedances are equal and unequal for odd and even order, respectively. Therefore, even-order Chebyshev filters possess impedance-transformation property, which may be exploited for certain impedance matching besides low-pass filtering, or requires a matching network if equal source and load impedances are desired.

2.2.2.3 Impedance and Frequency Scaling

In the low-pass filter prototypes, we assume that the source or load impedance and cutoff frequency are normalized ($g_o = 1$ and $\omega'_c = 1$ rad/s). In actual low-pass filters, however, both g_o and cutoff frequency are different from 1. This can be resolved by scaling the element values of the low-pass prototypes to the desired impedance and cutoff frequency of the actual filters.

Impedance Scaling
In the impedance scaling, we assume the source impedance is to be transferred from 1 (of the prototype) to R_s (of the actual filter). The impedance-scaled inductance, capacitance, resistance and conductance can be easily derived as:

$$
\begin{aligned}
L_k &= R_s L'_k & C_k &= \frac{C'_k}{R_s} \\
R &= R_s R' & G &= \frac{G'}{R_s}
\end{aligned}
\tag{2.3}
$$

where the primed and unprimed parameters are for the prototype and actual filters, respectively.

Frequency Scaling

In the frequency scaling, the cutoff frequency is transformed from $\omega'_c = 1$ for the prototype to $\omega_c \neq 1$ for the actual filter. The transformation is described as flows.

Low-Pass Filter Prototype	Low-Pass Filter
$\omega'_c = 1$	ω_c
ω'	$\frac{\omega}{\omega_c}$
$j\omega' L'_k$	$j\frac{\omega}{\omega_c} L'_k$
$j\omega' C'_k$	$j\frac{\omega}{\omega_c} C'_k$

The frequency-scaled inductance and capacitance can then be obtained as

$$L_k = \frac{L'_k}{\omega_c}$$

$$C_k = \frac{C'_k}{\omega_c} \tag{2.4}$$

Impedance and Frequency Scaling Combing the impedance and frequency scaling leads to the following results:

Low-Pass Filter Prototype	Low-Pass Filter
$R_o = 1\,\Omega,\ \omega'_c = 1\,\text{rad/s}$	$R_s,\ \omega_c$
$R',\ G'$	$R_s R',\ \frac{G'}{R_s}$
L'_k	$\frac{R_s L'_k}{\omega_c}$
C'_k	$\frac{C'_k}{\omega_c R_s}$

The low-pass filter elements after the impedance and frequency scaling are now given as

$$L_k = \frac{R_s L'_k}{\omega_c}$$

$$C_k = \frac{C'_k}{\omega_c R_s} \tag{2.5}$$

It is noted that the frequency transformation from ω' to ω/ω_c serves as the basic transformation from the low-pass filter prototype to a low-pass filter. The low-pass filter prototype can also be transferred to high-pass, band-stop and band-pass filters using appropriate frequency transformation to be discussed later.

2.2.2.4   Attenuation in Low-Pass Filters

The attenuation of an nth order maximally flat low-pass filter can be derived as

$$A_m = 1 + \alpha \left(\frac{\omega}{\omega_c}\right)^{2n} \tag{2.6}$$

where, again, n is the number of low-pass filter elements representing the order of the low-pass filter, ω_c is the cut-off frequency of the low-pass filter, and $\alpha = 10^{0.1\alpha_{dB}} - 1$ represents the maximum pass-band loss. For the maximally flat low-pass filters, the maximum insertion loss α or α_{dB} always occurs at the cutoff frequency. Another word, the cut-off frequency of the maximally flat low-pass filters is defined at the point corresponding to the maximum insertion loss. Typically, this point is chosen at 3-dB loss.

The attenuation response of an nth order Chebyshev low-pass filter is characterized by

$$A_c = \begin{cases} 1 + \alpha \cos^2\left[n \cos^{-1}\left(\frac{\omega}{\omega_c}\right)\right], & \omega \leq \omega_c \\ 1 + \alpha \cosh^2\left[n \cosh^{-1}\left(\frac{\omega}{\omega_c}\right)\right], & \omega \geq \omega_c \end{cases} \tag{2.7}$$

The maximum insertion loss α or α_{dB} does not necessarily occur at the cut-off frequency and is often referred to as the pass-band ripple for the Chebyshev filters. Figure 2.2 illustrates the Chebyshev low-pass filter's response.

Equations (2.6) and (2.7) can be used to calculate the attenuation of any type of filters provided that a suitable normalized frequency that maps the response of the low-pass filter prototype to that of the corresponding filter is used.

Fig. 2.2 Frequency response of Chebyshev low-pass filters. Reprinted, with permission, from [2]

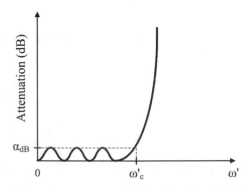

2.2.2.5 Low-Pass Filter Design

Lumped-element low-pass filters can be designed directly from the low-pass filter prototypes using the following procedure. First, the filter order or number of filter elements is determined from the filter specifications (maximum insertion loss, cut-off frequency, and stop-band rejection at a certain frequency) using the attenuation Eqs. (2.6) and (2.7) for the maximally flat and Chebyshev filters, respectively. Second, the element values g_k (k = 0, 1, 2, …, n + 1) of the low-pass filter prototype are calculated using (2.1) or (2.2). Third, the prototype values are converted to the values of the actual filter having $R_s \neq 1\,\Omega$, $\omega_c \neq 1$ rad/s using the foregoing impedance and frequency scaling equations. It is note that R_s is typically equal to 50 Ω.

2.3 High-Pass Filter

As mentioned earlier, the low-pass filter prototypes serve as the basis from which the responses of all filters can be derived through proper frequency transformations. We have seen the transfer from a low-pass filter prototype to an (actual) low-pass filter via the frequency transformation from ω' (of low-pass filter prototype) to ω/ω_c (of low-pass filter) where $\omega'_c = 1$ rad/s is the low-pass filter prototype's cutoff frequency.

The response of a high-pass filter can be obtained from that of a low-pass filter prototype by transforming ω' (of low-pass filter prototype) to ω (of high-pass filter) using the following form:

$$\frac{\omega'}{\omega'_c} = -\frac{\omega_c}{\omega} \tag{2.8}$$

where ω_c is the cutoff frequency of the high-pass filter. Applying this frequency transformation to the low-pass filter prototype, we can map its response onto another response as illustrated in Fig. 2.3, which indeed represents a high-pass filter. It is

Fig. 2.3 Frequency response of high-pass filters. Reprinted, with permission, from [2]

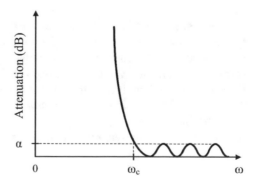

Fig. 2.4 Schematic of a high-pass filter. Reprinted, with permission, from [2]

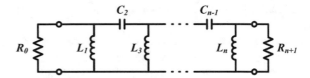

noted that ω' and ω, as defined in Figs. 2.2 and 2.3, respectively, are the radian frequency in the low-pass filter prototype and high-pass filter domains, respectively.

Applying the frequency mapping given in (2.8) to the inductors and capacitors making up the low-pass filter prototypes, we obtain the following transformations:

Low-Pass Filter Prototype High-Pass Filter

$$j\omega' L'_k \qquad\qquad -j\frac{\omega_c \omega'_c}{\omega} L'_k = \frac{1}{j\omega\left(\frac{1}{\omega_c \omega'_c L'_k}\right)} = \frac{1}{j\omega C_k}$$

$$\frac{1}{j\omega' C'_k} \qquad\qquad j\omega\left(\frac{1}{\omega_c \omega'_c C'_k}\right) = j\omega L_k$$

It is apparent from these transformations that the inductors and capacitors in the low-pass filter prototypes become the capacitors and inductors in high-pass filters, respectively, through the frequency mapping (2.8) as

$$C_k = \frac{1}{\omega_c \omega'_c L'_k}$$

$$L_k = \frac{1}{\omega_c \omega'_c C'_k} \tag{2.9}$$

It is recalled that $\omega'_c = 1$ rad/s for the low-pass filter prototypes. Figure 2.4 shows the schematic of a high-pass filter transformed from the low-pass filter prototype in Fig. 2.1a for n odd. Similarly, other high-pass filters can be obtained from other low-pass filter prototypes shown in Fig. 2.1 through the frequency mapping (2.8) and the transformation Eq. (2.9). The impedance scaling as described earlier in (2.3) for the low-pass filters is then applied to obtain the final element values for the high-pass filters.

The design of a high-pass filter can be proceeded from a low-pass filter prototype as follows. First, the number of filter elements is determined by calculating the normalized low-pass filter prototype frequency ω'/ω'_c given in the frequency transformation (2.8). This value is then used to determine the number of the low-pass filter elements, which is the same as that of the high-pass filter elements as done for the low-pass filter design. The attenuation characteristic of the high-pass filter can be determined from (2.6) and (2.7) for the maximally flat and Chebyshev topology, respectively, upon applying the frequency mapping (2.8).

2.4 Band-Pass Filter

The response of a band-pass filter can be obtained from that of a low-pass filter prototype by the transforming ω' (of low-pass filter prototype) to ω (of band-pass filter) according to

$$\frac{\omega'}{\omega_c'} = \frac{1}{\Delta\omega}\left(\frac{\omega}{\omega_o} - \frac{\omega_o}{\omega}\right) \tag{2.10}$$

where $\Delta\omega$ is the fractional bandwidth defined as

$$\Delta\omega = \frac{\omega_2 - \omega_1}{\omega_o} \tag{2.11}$$

and ω_o is the (design) center frequency defined as

$$\omega_o = \sqrt{\omega_1\omega_2} \tag{2.12}$$

with ω_1 and ω_2 being the lower and upper frequency of the pass band, respectively. Applying the frequency transformation according to (2.10) to the low-pass filter prototype, we can map its response onto another response as illustrated in Fig. 2.5, which characterizes a band-pass filter. ω' and ω, as defined in Fig. 2.5, are the radian frequency in the low-pass filter prototype and band-pass filter domains, respectively. It is noted that additional pass-bands at higher frequencies (harmonics) occur if distributed-element resonators are used due to the resonators' inherent multiple resonances. It is noted that the frequency mapping (2.10) may result in a negative value. This, however, does not cause any issue in the filter design, and hence the negative sign is omitted in the design. The possible negative sign is due to the mathematical mapping between the low-pass and band-pass filters. The band-pass frequency response below ω_o seen in Fig. 2.5 corresponds to the mirror of the low-pass filter response in the negative frequency region in Fig. 2.2. Since the responses of the low-pass filter in the positive frequency region and its mirror image in the negative

Fig. 2.5 Frequency response of band-pass filters. Reprinted, with permission, from [2]

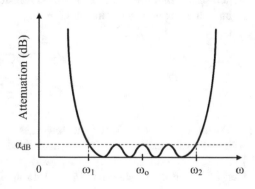

frequency region are identical with frequency shifting, the resultant band-pass filter response through the mapping is correct.

Applying the frequency mapping given in (2.10) to the series inductors of the low-pass filter prototypes, we can derive

$$
\begin{aligned}
j\omega' L'_k &= j\frac{\omega'_c}{\Delta\omega}\left(\frac{\omega}{\omega_o} - \frac{\omega_o}{\omega}\right)L'_k \\
&= j\omega\left(\frac{\omega'_c L'_k}{\omega_o \Delta\omega}\right) + \frac{1}{j\omega\left(\frac{\Delta\omega}{\omega'_c \omega_o L'_k}\right)} \\
&= j\omega L_k + \frac{1}{j\omega C_k}
\end{aligned}
\tag{2.13}
$$

which shows that the series inductor is transferred into a series resonator consisting of an inductor and a capacitor having

$$
\begin{aligned}
L_k &= \frac{\omega'_c L'_k}{\omega_o \Delta\omega} \\
C_k &= \frac{\Delta\omega}{\omega'_c \omega_o L'_k}
\end{aligned}
\tag{2.14}
$$

Similarly, we obtain for the shunt capacitors of the low-pass filter prototypes, after applying the frequency mapping:

$$
\begin{aligned}
j\omega' C'_k &= j\frac{\omega'_c}{\Delta\omega}\left(\frac{\omega}{\omega_o} - \frac{\omega_o}{\omega}\right)C'_k \\
&= j\omega\left(\frac{\omega'_c C'_k}{\omega_o \Delta\omega}\right) + \frac{1}{j\omega\left(\frac{\Delta\omega}{\omega'_c \omega_o C'_k}\right)} \\
&= j\omega C_k + \frac{1}{j\omega L_k}
\end{aligned}
\tag{2.15}
$$

which shows that the shunt capacitor is transferred into a shunt resonator consisting of an inductor and a capacitor having

$$
\begin{aligned}
L_k &= \frac{\Delta\omega}{\omega'_c \omega_o C'_k} \\
C_k &= \frac{\omega'_c C'_k}{\omega_o \Delta\omega}
\end{aligned}
\tag{2.16}
$$

Again, $\omega'_c = 1$ rad/s for the low-pass filter prototypes. Figure 2.6 shows the schematic of a band-pass filter transformed from the low-pass filter prototype in Fig. 2.1a for n odd. Similarly, other band-pass filters can be obtained from other low-pass filter prototypes shown in Fig. 2.1 through the frequency mapping (2.10) and the

Fig. 2.6 Schematic of a band-pass filter. Reprinted, with permission, from [2]

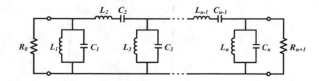

transformation Eqs. (2.14) and (2.16). The impedance scaling described earlier for the low-pass filters is then applied to obtain the final element values for the band-pass filters.

The design of a band-pass filter can be procceded from a low-pass filter proto-type as follows. First, the number of filter elements is determined by calculating the normalized low-pass filter prototype frequency ω'/ω'_c given in the frequency trans-formation (2.10). This value is then used to determine the number of the low-pass filter elements which is the same as that of the band-pass filter elements as done for the low-pass filter design. The attenuation charactcristic of the band-pass filter can be determined from (2.6) and (2.7) for the maximally flat and Chebyshev topology, respectively, upon using the frequency mapping (2.10).

2.5 Band-Stop Filter

The response of a band-stop filter can be obtained from that of a low-pass filter prototype by transforming ω' (of low-pass filter prototype) to ω (of band-stop filter) using the following form:

$$\frac{\omega'_c}{\omega'} = \frac{1}{\Delta\omega}\left(\frac{\omega}{\omega_o} - \frac{\omega_o}{\omega}\right) \tag{2.17}$$

where $\Delta\omega$ is the fractional bandwidth defined as

$$\Delta\omega = \frac{\omega_2 - \omega_1}{\omega_o} \tag{2.18}$$

and ω_o is the center frequency defined by

$$\omega_o = \sqrt{\omega_1\omega_2} \tag{2.19}$$

with ω_1 and ω_2 being the lower and upper frequency of the stop band at the maxi-mum insertion loss or pass-band ripple (α_{dB}), respectively. Applying this frequency transformation to the low-pass filter prototype, we can map its response onto the response of a band-stop filter as illustrated in Fig. 2.7. ω, as defined in Fig. 2.7, is the radian frequency in the band-stop filter domain.

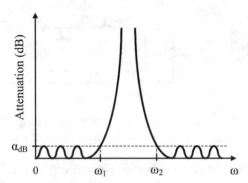

Fig. 2.7 Frequency response of band-stop filters. Reprinted, with permission, from [2]

Applying the frequency mapping (2.17) to the series inductors of the low-pass filter prototypes, we obtain

$$
\begin{aligned}
\frac{1}{j\omega' L_k'} &= \frac{1}{j\omega_c' \Delta\omega}\left(\frac{\omega}{\omega_o} - \frac{\omega_o}{\omega}\right)\frac{1}{L_k'} \\
&= j\frac{\omega_o}{\omega\omega_c' \Delta\omega L_k'} + \frac{\omega}{j\omega_o\omega_c' \Delta\omega L_k'} \\
&= j\omega C_k + \frac{1}{j\omega L_k}
\end{aligned}
\tag{2.20}
$$

which leads to

$$
\begin{aligned}
\omega L_k &= \frac{\omega_o\omega_c' \Delta\omega L_k'}{\omega} \\
\omega C_k &= \frac{\omega_o}{\omega\omega_c' \Delta\omega L_k'}
\end{aligned}
\tag{2.21}
$$

from which, we obtain at the center frequency ω_o:

$$
\begin{aligned}
L_k &= \frac{\omega_c' \Delta\omega L_k'}{\omega_o} \\
C_k &= \frac{1}{\omega_c'\omega_o \Delta\omega L_k'}
\end{aligned}
\tag{2.22}
$$

It is now apparent that each series inductor is transferred into a parallel resonator consisting of an inductor and a capacitor whose respective inductance and capacitance are given in (2.22).

Similarly, we can obtain for the shunt capacitors of the low-pass filter prototypes, after applying the frequency mapping (2.17):

Fig. 2.8 Schematic of a band-stop filter. Reprinted, with permission, from [2]

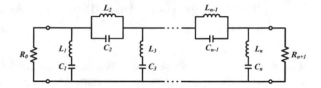

$$\frac{1}{j\omega'C_k'} = \frac{1}{j\omega_c'\Delta\omega}\left(\frac{\omega}{\omega_o} - \frac{\omega_o}{\omega}\right)\frac{1}{C_k'}$$

$$- j\frac{\omega_o}{\omega\omega_c'\Delta\omega C_k'} + \frac{\omega}{j\omega_o\omega_c'\Delta\omega C_k'}$$

$$= j\omega L_k + \frac{1}{j\omega C_k} \qquad (2.23)$$

from which, we get

$$\omega L_k = \frac{\omega_o}{\omega\omega_c'\Delta\omega C_k'}$$

$$\omega C_k = \frac{\omega_o\omega_c'\Delta\omega C_k'}{\omega} \qquad (2.24)$$

where $\omega_c' = 1$ rad/s. A shunt capacitor is thus transferred into a series resonator consisting of an inductor and a capacitor having

$$L_k = \frac{1}{\omega_o\omega_c'\Delta\omega C_k'}$$

$$C_k = \frac{\omega_c'\Delta\omega C_k'}{\omega_o} \qquad (2.25)$$

Figure 2.8 shows the schematic of a band-stop filter transformed from the low-pass filter prototype in Fig. 2.1a for n odd. Similarly, other band-stop filters can be obtained from other low-pass filter prototypes shown in Fig. 2.1 through the frequency mapping (2.17) and the transformation Eqs. (2.22) and (2.25). The impedance scaling described earlier for the low-pass filter is finally applied to obtain the final element values for the band-stop filter.

The design of a band-stop filter can be proceeded from a low-pass filter proto-type as follows. First, the number of filter elements is determined by calculating the normalized low-pass filter prototype frequency ω'/ω_c' given in the frequency trans-formation (2.17). This value is then used to determine the number of the low-pass filter elements which is the same as that of the band-stop filter elements as done for the low-pass filter design. The attenuation characteristic of the band-stop filter can be determined from (2.6) and (2.7) for the maximally flat and Chebyshev topology, respectively, upon using the frequency mapping (2.17).

2.6   Filter Design Using Impedance and Admittance Inverters

It is known that a series resonator can be transformed into a parallel resonator or vice versa by using an impedance or admittance inverter. Impedance and admittance inverters can therefore facilitate the design of band-pass and band-stop filters, allowing only one kind of resonators (either series or parallel resonator), instead of both types of resonators, to be used. We will illustrate this design approach using band-pass filters. Similar formulation can be employed for band-stop filter design using impedance and admittance inverters.

As with a typical filter design, we start with a low-pass filter prototype such as that in Fig. 2.1a for n odd. This low-pass filter prototype can be modified to include the impedance inverters as shown in Fig. 2.9a. Applying the low-pass to band-pass frequency transformation given in (2.10), we replace the series inductors by series resonators to obtain the band-pass filter in Fig. 2.9b. In addition to the impedance inverters, admittance inverters can also be used to realize band-pass filters as shown in Fig. 2.10. It is noted that some band-pass filters (or band-stop filters) lend

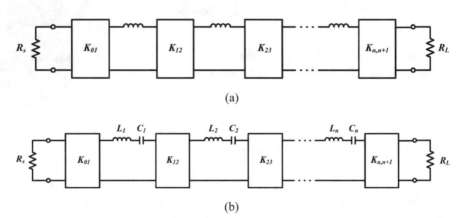

(a)

(b)

Fig. 2.9 a Modified low-pass filter prototype using impedance inverters. **b** Band-pass filter consisting of only series resonators of inductors and capacitors and impedance inverters. K_{ok}, $k = 1$, 2, ..., $n + 1$, are the impedance inverter parameters. R_S and R_L are the source and load impedance, respectively. Reprinted, with permission, from [2]

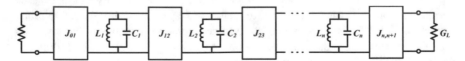

Fig. 2.10 Band-pass filter consisting of only shunt resonators of inductors and capacitors and admittance inverters. J_{ok}, $k = 1, 2$, ..., $n + 1$, are the admittance inverter parameters. $G_S = 1/R_S$ and $G_L = 1/R_L$ are the source and load conductance, respectively. Reprinted, with permission, from [2]

themselves well to the impedance or admittance inverter, and so either the impedance or admittance inverter can be used depending on the band-pass filter (or band-stop filter) topology.

2.7 Multiband Band-Pass Filters

Multiband band-pass filters can be designed by incorporating multiband resonators into (single-band) band-pass filters. This approach can also be used to design multiband band-stop filters. The design basically starts with a low-pass filter prototype and transform it into a single-band band-pass filter, which is then transformed into a dual-band band-pass filter and its counterpart using admittance inverters (J-inverters). For simplicity, without loss of generality, we will illustrate the design of multiband band-pass filters with a dual-band band-pass filter designed from a 3rd order (single-band) band-pass filter.

We begin with a 3rd order low-pass filter prototype as shown in Fig. 2.11a, obtained from Fig. 2.1 with inductors as the first and last elements. The low-pass filter prototype g_0, g_1, g_2, g_3 and g_4 are determined using Eqs. (2.1) or (2.2) for maximally flat or Chebyshev response, respectively. Following the procedure described in Sec. 2.4, the

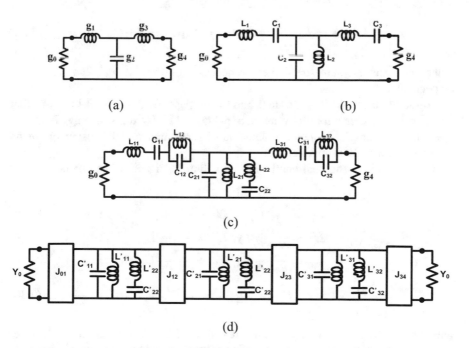

Fig. 2.11 Low-pass filter prototype (**a**) and its transformation into single-band band-pass filter (**b**), dual-band band-pass filter (**c**), and dual-band band-pass filter with admittance inverters (**d**)

low-pass filter prototype is transferred into a band-pass filter shown in Fig. 2.11b, whose elements are obtained from (2.14) and (2.16), with $\omega'_c = 1$ rad/s (for the low-pass filter prototype) and R_s being the (actual) source impedance of the filter, as

$$
\begin{aligned}
L_1 &= \frac{g_1 R_s}{\omega_o \Delta\omega} & C_1 &= \frac{\Delta\omega}{\omega_o g_1 R_s} \\
L_2 &= \frac{\Delta\omega R_s}{\omega_o g_2} & C_2 &= \frac{g_2}{\omega_o \Delta\omega R_s} \\
L_3 &= \frac{g_3 R_s}{\omega_o \Delta\omega} & C_3 &= \frac{\Delta\omega}{\omega_o g_3 R_s}
\end{aligned}
\tag{2.26}
$$

where ω_o and $\Delta\omega$ are, again, the (design) center frequency and fractional bandwidth of the band-pass filter. A dual-band band-pass filter can now be derived following [3, 4].

Replacing the inductor and capacitor of the single-band band-pass filer as shown in Fig. 2.11b with a series and parallel resonator, respectively, transforms it into a dual-band band-pass filter, as shown in Fig. 2.11c. The dual-band band-pass filter consists of series and parallel resonators in each branch, whose values can be determined from

$$
\begin{aligned}
L_{11} &= \frac{g_1 R_s}{\Delta\omega(\omega_{o2}-\omega_{o1})} & L_{12} &= \frac{g_1(\omega_{o2}-\omega_{o1})R_s}{\omega_o^2 \Delta\omega} \\
L_{21} &= \frac{\Delta\omega(\omega_{o2}-\omega_{o1})}{g_2 \omega_o^2 R_s} & L_{22} &= \frac{\Delta\omega}{g_2(\omega_{o2}-\omega_{o1})} \\
L_{31} &= \frac{g_3 R_s}{\Delta\omega(\omega_{o2}-\omega_{o1})} & L_{32} &= \frac{g_3(\omega_{o2}-\omega_{o1})R_s}{\omega_o^2 \Delta\omega}
\end{aligned}
\tag{2.27}
$$

$$
L_{11}C_{11} = L_{12}C_{12} = L_{21}C_{21} = L_{22}C_{22} = L_{31}C_{31} = L_{32}C_{32} = \frac{1}{\omega_o^2} \tag{2.28}
$$

where ω_{o1} and ω_{o2} are the center frequencies of the 1st and 2nd pass-band, respectively.

To facilitate the design, the dual-band band-pass filter in Fig. 2.11c is further changed into another topology as shown Fig. 2.11d by transforming the series branches into parallel branches consisting of series and parallel resonators using admittance inverters (J-inverters).

The elements of the dual-band band-pass filter in Fig. 2.11d can be determined from

$$
\begin{aligned}
L'_{11} &= \frac{g_0 G_s}{J_{01}^2} C_{11} & L'_{12} &= \frac{g_0 G_s}{J_{01}^2} C_{12} \\
L'_{21} &= \frac{J_{12}^2}{g_0 G_s J_{12}^2} L_{21} & L'_{22} &= \frac{J_{12}^2}{g_0 G_s J_{12}^2} L_{22} \\
L'_{31} &= \frac{J_{12}^2}{g_0 G_s J_{23}^2} L_{31} & L'_{32} &= \frac{J_{12}^2}{g_0 G_s J_{23}^2} L_{32}
\end{aligned}
\tag{2.29}
$$

$$
L'_{11}C'_{11} = L'_{12}C'_{12} = L'_{21}C'_{21} = L'_{22}C'_{22} = L'_{31}C'_{31} = L'_{32}C'_{32} = \frac{1}{\omega_o^2} \tag{2.30}
$$

where $G_s = 1/R_s$. Each of the parallel branches of this dual-band band-pass filter works as a dual-band resonator resonating at two distinct frequencies ω_{o1} and ω_{o2}, to be discussed later, making the filter a band-pass filter working concurrently at two separate passbands centered at ω_{o1} and ω_{o2}.

Admittance inverters (and their impedance-inverter counterparts) can be used to transform series resonators into parallel resonators, impedances to admittances, series circuits into parallel circuits, or vice versa. They have been used widely to facilitate analysis and design of filters by transferring a filter topology to another one that is more suitable for analysis and design.

The most basic admittance inverter is a quarter-wavelength transmission line [1, 2]. Within a narrow frequency range around the quarter-wavelength frequency, the impedance-admittance inversion function of the admittance inverter is almost exact. An ideal admittance inverter with admittance parameter J is defined as a network that functions as a quarter-wavelength transmission line having a constant characteristic admittance J at all frequencies. A quarter-wavelength transmission line forms the basic (ideal) inverter, from which other inverters can be based on. Any RF network can be considered as admittance (or impedance) inverter if it was designed to electrically behave as a quarter-wavelength transmission line; i.e., the image phase or phase of the transmission coefficient is an odd multiple of $\pm 90°$ and the image admittance (or impedance) is real in the operating frequency band. Using this concept, various RF networks can be configured and designed as inverters.

Figure 2.12a shows an ideal quarter-wavelength admittance inverter. Figure 2.12b, c show two lumped-element admittance inverters equivalent to the quarter-wavelength admittance inverter, which contain series positive and shunt negative capacitors and inductors [1]. The admittance parameter J of the inverters in Fig. 2.12a–c is Y_o, which is the characteristic admittance of the quarter-wavelength transmission line, ωC_j, $1/\omega L_j$, respectively.

The inverters in Fig. 2.12b, c contain shunt negative capacitance and inductance, respectively, and are suitable for use with parallel resonators that can absorb those negative elements to form circuits with only positive capacitance and inductance. For instance, the negative capacitances in the inverter in Fig. 2.12b can be absorbed into the positive capacitances of parallel resonators preceding and following the inverter, provided that the (positive) capacitances are larger than the (negative) capacitances.

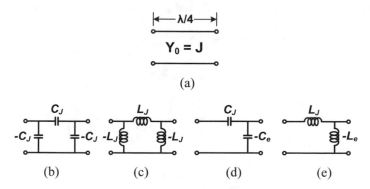

(a)

(b)   (c)   (d)   (e)

Fig. 2.12 Quarter-wavelength admittance inverter (**a**) and equivalent lumped-element admittance inverters (**b–e**)

Fig. 2.13 A dual-band
parallel resonator

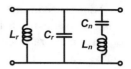

The admittance inverters in Fig. 2.12b, c, however, are not suitable for the first and last admittance inverters due to the difficulty in absorbing the negative inductors and capacitors at the source and load end of a filter. To overcome this problem, the inverters in Fig. 2.12b, c are transferred into other lumped-element inverters as shown in Fig. 2.12d, f, respectively, in which the inductance L_e and capacitance C_e can be derived as

$$L_e = \frac{1 + (\omega_o L_J Y_o)^2}{\omega_o^2 L_J Y_o^2} \tag{2.31}$$

$$C_e = \frac{C_J}{1 + \left(\frac{\omega_o C_J}{Y_o}\right)^2} \tag{2.32}$$

where ω_o is the design frequency and Y_o is the terminating admittance. The negative inductance C_e and capacitance C_e of these inverters can be conveniently absorbed into the adjacent inductor and capacitor of the preceding or following resonator. Selection of a lumped-element admittance inverter depends not only on possibility of absorbing the negative elements into the preceding or following resonator, but also on desired filtering characteristics (low or high pass response) and possibility and simplicity of layout.

In general, a circuit would function over multiple bands if its components can work over such multi-band simultaneously. Specifically, a dual-band band-pass filter would need to employ dual-band resonators functioning resonance structures over these two bands. Figure 2.13 shows a dual-band resonator operating at two distinctive frequencies, whose elements C_n, L_n, C_r and L_r can be calculated as [5]

$$C_n = 2\Delta\omega_s / Z_o \omega_s \tag{2.33}$$

$$L_n = 1 / \omega_s^2 C_n \tag{2.34}$$

$$C_r = \frac{1}{\left[\omega_{o1}^2 + \omega_{o2}^2 - \omega_s^2 - \frac{(\omega_{o1}^2 + \omega_{o2}^2)^2 - (\omega_{o2}^2 - \omega_{o1}^2)^2}{4\omega_s^2}\right] L_n} \tag{2.35}$$

$$L_r = \frac{1}{\left[\omega_{o1}^2 + \omega_{o2}^2 - \omega_s^2 - \frac{1}{L_n C_r}\right] C_r} \tag{2.36}$$

where Z_o is the terminating impedance, ω_s is the stop-band center frequency, $\Delta\omega_s$ is the stop-band fractional bandwidth, and ω_{o1} and ω_{o2} are the 1st and 2nd pass-band center frequency, respectively.

This dual-band resonator, as can be seen, resembles each of the parallel branches of the band-pass filter in Fig. 2.11d. This band-pass filter therefore can perform as a dual-band band-pass filter and its design can be readily done using the foregoing equations and desired filter specifications. It in fact provides a basis for the design of various dual-band filter-switches presented in this book.

2.8 Summary

This chapter covers the fundamentals of low-pass, high-pass, band-pass, and band-stop filters. It also discusses dual-band band-pass filters employing dual-band resonators, which form the basis for the design of the dual-band band-pass filter-switches presented in this book.

References

1. Matthaei GL, Young L, Jones EMT (1980) Microwave filters, impedance-matching networks, and coupling structures. Artech House, Dedham, MA
2. Nguyen C (2015) Radio-frequency integrated-circuit engineering. Wiley, New York
3. Guan X, Ma Z, Cai P, Kobayashi Y, Anada T, Hagiwara G (2005) A dual-band bandpass filter synthesized by using frequency transformation and circuit conversion technique. In: Proceedings of the IEEE Asia-Pacific microwave conference (APMC '05), Yokohama, Japan, vol 4, Dec 2005
4. Guan X, Ma Z, Cai P, Kobayashi Y, Anada T, Hagiwara G (2006) Synthesis of dual-band bandpass filters using successive frequency transformations and circuit conversions. IEEE Microw Wirel Compon Lett 16(3):110 112
5. Mao S-G, Wu M-S (2008) Design of artificial lumped-element coplannar waveguide filters with controllable dual-passband responses. IEEE Trans Microw Theory Tech 56(7):1684–1692

Chapter 3
Switches

3.1 Introduction

Switches are important components found in many RF systems from single-pole single-throw (SPST) to multi-pole multi-throw switches. They can be used as a stand-alone component or integrated within subsystems or systems—for instance, transmit/receive (T/R) switches used in systems to accommodate the operation of transmitters and receivers sharing a common antenna. Switches are also used as an integral part of a component to enable the component to perform a particular function such as a SPST switch in a RF pulse-former component to modulate a continuous-wave (CW) signal to form a RF pulse signal. More complex multi-pole multi-throw switches, such as single-pole double-throw (SPDT) or transmit/receive (T/R) switches, can be employed to achieve various system functions like transmission and reception of dual-polarized signals using a single dual-polarized antenna. Besides the fundamental function of switching seen in commonly used switches, other functions such as filtering and attenuation can also be incorporated into a switch to form a multi-function switch like a filter switch or a switching attenuator. RF switches can be employed in Time Division Duplexing (TDD) systems as T/R switches to facilitate switching between the common antenna and the transmitter and receiver. In Frequency Division Duplexing (FDD) systems, RF switches can be implemented with branches operating at different frequency bands.

This chapter presents the fundamentals of switches including insertion loss, isolation, power handling and nonlinearity, and figure of merit. Moreover, SPST, SPDT, and T/R switches are also addressed.

© The Author(s) 2020
C. Nguyen and Y. Um, *Multiband Dual-Function CMOS RFIC Filter-Switches*,
SpringerBriefs in Electrical and Computer Engineering,
https://doi.org/10.1007/978-3-030-46248-2_3

3.2 Fundamentals of Switches

3.2.1 Switch Operation

Switches function based on the on- and off-sate of the employed semiconductor devices (diodes or transistors). For discussion purposes without loss of generality, we will use MOSFET. At low frequencies, MOSFET exhibits a very small resistance between the drain and source terminals when the dc biased voltage applied to the gate-source is higher than the threshold voltage (on state) and a very large resistance when the gate-source voltage is lower than the threshold voltage (off state). It is this difference in the impedances between the on- and off-state that the switch operation relies on. The function and performance of switches are based on the difference in the reflection of signals instead of the dissipation of signals. As MOSFETs and hence switches are typically operated as a passive device, there is no dc power dissipated in switches. The switch operation is therefore dependent upon the operation or, specifically, passive models of MOSFET under its on- and off-state. These passive models are covered in Chap. 4.

The simplest switch type is SPST, which is operated as on and off states between the switch's two ports. In more complex switches, the operation is typically based on the on-state between the input port and one of the output ports while the paths between the input and the remaining output ports are turned off. For instance, a T/R switch, or equivalently a single-pole double-throw (SPDT) switch, operates under on-state between the antenna port and the transmitter port and off-state between the antenna and receiver ports in the transmission mode while, in the reception mode, its antenna-receiver port is on and the antenna-transmitter port is off.

Switches are formed by transistors connected in series only, shunt only or alternating series and shunt in single or multiple paths. Therefore, a series transistor, shunt transistor, or combined series and shunt transistors serve as the basic building element for various switches. Among various types of switches, the SPST switch having a single path between the input and output ports represents a basic element found in other switches with multi-path like SPDT switches having two paths between the input and two output ports. As such, the analysis and performance of a series transistor, shunt transistor, or combination of series and shunt transistors and SPST switches, as well the design of SPST switches, form the foundation for the analysis and design of switches. For illustration of the fundamentals of switches, while retaining simplicity without loss of generality, we consider a very basic switch modeled with a simple on-resistance R_{on} and off-capacitance C_{off}, as shown in Fig. 3.1, which typically represents a SPST switch consisting of a series-connected MOSFET.

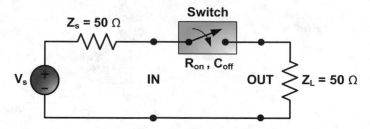

Fig. 3.1 Basic switch model. The source (Z_s) and load (Z_L) are assumed to be 50 Ω as typical

3.2.2 Insertion Loss

Typical switches are operated as passive devices and, hence, suffer losses. This loss, typically known as insertion loss, just like other passive components such as filters, is important for the use of switches in systems. This importance is more pronounced when a switch is used in front of a low-noise amplifier (LNA) such as a T/R switch, as the receiver noise figure is increased by an amount approximately equal to the insertion loss of the switch. This is especially important in the millimeter-wave regime, where the insertion loss of a millimeter-wave switch is relatively high while high gain and low noise figure of a millimeter-wave LNA are difficult to obtain. It is noted that designing a switch to operate in an active mode or including a gain stage (acting as an amplifier) in a switch to overcome the loss, while may improve the performance in certain use such as for transmitters, would degrade the performance for receivers due to increased noise figure.

The insertion loss (IL) (in dB) between the input and output ports of the switch occurred in its on-state can be calculated as

$$IL = -20 \log_{10} \left| \frac{Z_L + Z_s}{Z_L + Z_s + R_{on}} \right| \tag{3.1}$$

For practical on-state models consisting of resistors, capacitors and inductors, R_{on} should be replaced with the actual impedance of the on-state model as in [1]. As seen in (3.1), R_{on} should be as low as possible for minimum insertion loss. This insertion loss, particular, affects the operation of a transmitter and receiver, connected immediately before or after a switch (e.g., in a T/R switch). Higher loss leads to lower power that can be radiated from a transmitter through an antenna and higher noise figure (NF) for a receiver as seen in the following. We consider a switch and low-noise amplifier (LNA) in a typical receiver chain as shown in Fig. 3.2.

The total NF (in dB) of the receiver chain is given as

$$F = IL + \frac{F_{LNA} - 1}{IL^{-1}} + \cdots = IL \cdot F_{LNA} + \cdots \tag{3.2}$$

Fig. 3.2 SPST switch and
LNA in a typical RF receiver
chain

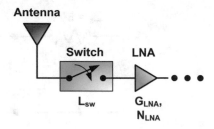

where IL and F_{LNA} represent the insertion loss (in dB) and NF (in dB) of the switch and LNA, respectively. As can be seen, the NF of the receiver is directly affected by the switch's insertion loss and, to reduce the NF of the receiver, the insertion loss of the switch should be as low as possible.

3.2.3 Isolation

Isolation in switches measures how well the two ports in a switch are isolated from each other when the switch between these ports is off. In the design of switches, normally, it is relatively difficult to achieve high isolation than low insertion loss. Moreover, in some applications, isolation is more important than insertion loss. For instance, in a T/R switch, the isolation between the transmitter port and antenna port (when the switch between the transmitter and antenna is turned off) and the isolation between the receiver port and antenna port (when the switch between the receiver and antenna is turned off) are normally more important than the insertion loss, particularly the isolation between the transmitter and antenna since, while the loss may be overcome by increasing the gain or output power of the power amplifier, poor isolation causes large signal leakage from the transmitter to receiver, resulting in undesired effects.

Consider again a simple switch as shown in Fig. 3.1. The isolation (ISO) (in dB) between the input and output ports of the switch occurred in its off-state can be calculated using the same Eq. (3.1) as

$$ISO = -20 \log_{10} \left| \frac{Z_L + Z_s}{Z_L + Z_s + \left(j\omega C_{off} \right)^{-1}} \right| \tag{3.3}$$

A seen in (3.3), C_{off} should be as low as possible for maximum isolation. For practical off-state models consisting of capacitors, inductors and resistors, C_{off} should be replaced with the actual impedance of the off-state model as in [1].

It is particularly noted that the isolation of a switch preceding a LNA in a receiver chain is not considered in the characterization of the receiver's noise figure because the receiver is supposed not to function when the switch is switched off. However, in practical systems, a receiver does not stand alone by itself, but is connected to other subsystems, notably a transmitter, through switches, and hence the isolation of those switches in off-state could affect the noise figure of a receiver in operation. For instance, in a transceiver module, a T/R switch is typically used between the power amplifier (PA) and LNA of a transmitter and receiver, respectively. When the receiving switch connecting to the LNA is "on" for the receiver to operate, the transmitting switch connecting to the power amplifier is "off." Under this operation, the isolation of the transmitting off-switch would indirectly affect the noise figure of the receiver. A poor isolation would allow the noise of the transmitter to be injected into the receiver and degrading its overall noise figure.

3.2.4 Switching Time

Switching time is important for switches requiring high-speed switching. The switching speed of a MOSFET switch depends not only on the MOSFET itself, but also on the circuit environment in which the MOSFET is embedded. For instance, the gate resistor typically used for dc biasing at the gate affects the switching time and should be optimized to reduce the (RC) time constant, and hence resulting in a faster speed, while keeping a minimum effect on the insertion loss. The switching time can be determined by simulating or measuring the RF pulse of a continuous wave (CW) signal produced by turning the switch on and off. The rising and falling times between 10 and 90% of the maximum voltage are typically used for the switching time. These times are normally different for practical switches.

3.2.5 Power Handling Capability and Linearity

RF power handling capability and linearity are important parameters for switches, particularly when they are used in place where high RF power exists such as in the path connecting an antenna and a transmitter in a T/R switch or in environments where strong signals may bombard on antennas connecting to receivers. Due to the high power typically used in transmitters, it is crucial that the T/R switch can handle high RF power and has high linearity. It is important to note that switches are typically designed to operate in their passive state, which is essentially linear state, to maintain the purity of transmitting or receiving signals, except a reduction in signal amplitude. Nonlinear behavior, however, occurs with signals of sufficiently large power, which causes undesirable intermodulation products at different frequencies to occur, and needs to be avoided. The switch performance needs to be as strong and linear as possible in order to produce strong high-quality signal. The linearity measures the quality of switches in handling RF signals or, specifically, it measures the ability of switches to operate without distorting RF signals at high input power levels. The important aspect in signal or power handling is the maximum tolerable level of the input RF signal, which is described by a power compression magnitude, typically

1 dB. As the RF signal is increased beyond the linear range, reaching a large-signal level, the insertion loss is no longer constant, and the output power begins to saturate and then reduces as the input power is increased, causing switches to compress. The 1-dB power compression point (P_{1dB}) characterizes the power handling capability and measures the departure of switches from their linear operation in which the output signal power increases linearly with the input RF signal power. The 1-dB input power compression is the input RF signal level at which the actual output RF power is 1 dB less than the linearly increased output RF level. The 1-dB input power compression point is typically used as the maximum input power that a switch can handle. The linearity of switches is typically characterized by the third-order intercept point (IP_3).

Figure 3.3a shows the output power and insertion loss (or gain) as a function of the input power of a device under test (DUT) (herein, switch), from which the input (IP_{1dB}) and output (OP_{1dB}) 1-dB power compression points are defined. Herein, P_{1dB} specifies the loss 1-dB compression point. As the power level of the fundamental signal (or tone) injected to the DUT increases, the DUT maintains linear output power and constant loss (or gain). However, the output power and loss (or gain) start to deviate from the linearity and increase (or reduce) at a certain input power level, respectively. The P_{1dB} specifies the RF power at which the output power is 1-dB below the linearly increased output RF level and/or the loss (or gain) is 1-dB above (or below) the constant loss (or gain).

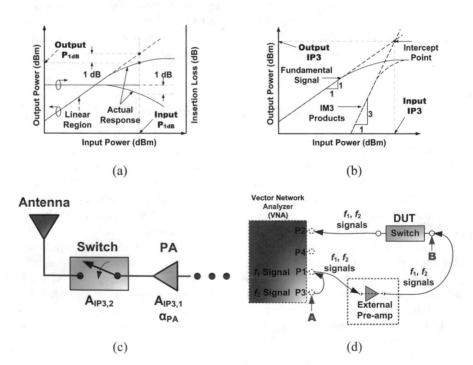

Fig. 3.3 Descriptions of **a** P_{1dB} and **b** IP_3. **c** Cascaded switch-PA in a transmitter. **d** Measurement set-up for IP3 of DUT (switch) with dual-tone signals at f_1 and f_2

Figure 3.3b shows the output power of the fundamental signal and third-order intermodulation (IM3) product as a function of the input power of a device under test (DUT) (herein, switch), from which the input (IIP3) and output (OIP3) third-order intercept points are defined. When two closely spaced fundamental tones (f_1 and f_2) are injected into a DUT, the DUT generates additional signals due to inter-modulation products because of its nonlinearity. Among the products, the IM3 products such as the signals at $2f_1$-f_2 and $2f_2$-f_1 can be located close to, and hence distort, the fundamental tones, and are thus needed to be considered in the analysis, design, and measurement of the DUT. As the fundamental tones' power level increases, the IM3 products' power levels also increase by three times of the fundamental signal's increment, as shown in Fig. 3.3b. At a certain input or output power level, the linearly extracted fundamental and IM3 signals' power levels are the same, at which the IP3 points are defined. The IIP3 and OIP3 power levels characterizes highest linearity levels of the DUT—beyond that, the DUT no longer functions as a linear device. When a switch, such as that in a T/R switch, is located after the PA in a transmitter, as shown in Fig. 3.3c, where high power is expected, the linearity of the switch becomes more critical. Consequently, the power handling capability of the switch should be as high as possible. The need of high IP3 for a switch used after a PA as shown in Fig. 3.3c is further confirmed from

$$\frac{1}{A_{IP3}^2} \approx \frac{1}{A_{IP3,1}^2} + \frac{\alpha_1^2}{A_{IP3,2}^2} \tag{3.4}$$

where A_{IP3} is the magnitude of the total IP$_3$ of the cascaded switch-PA; $A_{IP3,1}$ and $A_{IP3,2}$ are the IP$_3$ magnitudes of the PA and switch, respectively; and α_{PA}^2 is the gain of the PA. Equation (3.4) shows that, for a given PA, the higher the IP3 for the switch, the better IP3 for the transmitter.

Figure 3.3d shows a two-tone (f_1 and f_2) measurement setup for IP$_3$ using a vector network analyzer (VNA) capable of producing two internal sources at different frequencies. To generate the two-tone signals, the f_2 signal from Port 3 is combined with the f_1 signal at Port 1 through the internal combiner in the VNA as seen in Fig. 3.3d. The combined signal (f_1 and f_2) is then injected to the DUT (herein, switch). While the input power level injected to the switch at point 'B' could reach a level higher than the anticipated power handling capability of the switch, the output power level from the instrument at 'A' may be limited, especially at millimeter-wave frequencies. In order to increase the input power level at 'B' for the DUT, an additional external pre-amp is normally used as shown in Fig. 3.3d.

One important remark needs to be made at this point is that switches (and in fact, most components involving active devices) should be designed, simulated and measured under the same conditions at which they are intended to be operated since, in general, the active devices and design techniques used for the switches are affected by the operating conditions. For instance, for switches intended for high-power applications such as T/R switches, the performance of a switch such as insertion loss and isolation needs to be simulated and measured under both small- and large-signal conditions.

3.2.6 Figure of Merit

As for other RF components, the Figure of Merit (FOM) of a switch can be used as a parameter to evaluate the overall performance or, in another word, the quality of the switch. In principle, FOM can be defined in various forms involving some or all of the parameters characterizing important performances of a component. In the simplest way, a FOM of a switch consisting of a series MOSFET can be defined using only the two most fundamental switching parameters of MOSFET (on-resistance R_{on} and off-capacitance C_{off}) as

$$FoM = \frac{1}{R_{on}C_{off}} \tag{3.5}$$

Equation (3.5) shows that high FOM can be obtained with small R_{on}, corresponding to insertion loss, and small C_{off} corresponding to high isolation. This simple FOM defines the performance of a switch not based on its actual performance, but on the quality of the transistors employed in the switch. Other FOM's based on actual performance can be defined in terms of insertion loss (IL), isolation (ISO), P1dB, IP3, center frequency (f_o), and/or operating frequency range, etc., for instance,

$$FoM(dB) = ISO - IL + 20\log(f_o) \tag{3.6}$$

where FoM (dB) represents FoM in dB, and IL and ISO are assumed to be positive (in dB).

3.3 SPST Switches

SPST switch is the basic element for all switches and hence plays an important part in the design, analysis and performance of switches. Practical MOSFETs have finite impedances under their on- and off-state, which limit the amount of isolation that can be obtained. These finite impedances make it difficult to achieve high isolation using a single MOSFET, particularly in the high RF range where MOSFET's parasitics such as inductors have large reactances. In order to achieve high isolation or isolation over a wide bandwidth, multiple MOSFETs, placing next to each other or far apart, are needed.

It is noted that, to maintain symmetry in circuit layouts, which is crucial for some circuits, especially at millimeter-wave frequencies, use of two parallel transistors at the same location may be needed, and the use of symmetrical transmission lines such as coplanar waveguide (CPW) facilitates this kind of connection. Simple SPST switches can employ only a series MOSFET, a shunt MOSFET, and a pair of series/shunt MOSFETs. Details of SPST switches can be found in [1].

3.4 SPDT and T/R Switches

SPDT switches are one of the most widely used switches in RF systems, particularly as transmit/receive (T/R) switches for transceivers connecting to a single antenna. A SPDT switch has one input and two output branches and, in operation, one output branch is turned on while the other output branch is turned off. A SPDT switch hence requires two SPST switches effectively connected in parallel, one in each output branch with at least one MOSFET, to function. As such, the design of SDT switches is based on that for SPST switches and can consist of series, shunt, or combined series and shunt MOSFETs in each output arm. The filter-switches presented in this book are based on SPDT switches.

SPDT switches consist of two branches, each typically consisting of a SPST switch, and work based on the principle that one branch is on while the other is off. Figure 3.4 shows two fundamental SPDT switch topologies and their equivalent-circuit models with MOSFET's on-resistance and off-capacitance. The SPDT in Fig. 3.4a consists of two series MOSFETs, one in each output branch. One of the MOSFETs is turned on whereas the other is turned off simultaneously to allow a RF signal to pass from the input (Port 1) to the output (Port 2 or 3) corresponding to the

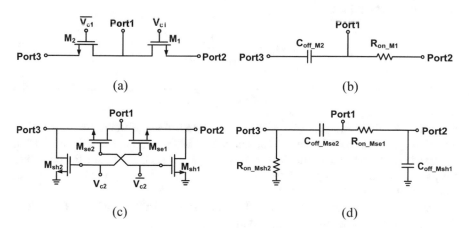

Fig. 3.4 SPDT switch employing series MOSFETs **a** and **b** its equivalent circuit when $V_{c1} = 1.8$ V, and **c** series-shunt MOSFETs and **d** its equivalent circuit when $V_{c2} = 1.8$ V

on-MOSFET. The other SPDT in Fig. 3.4c employs a series and shunt MOSFET in each arm. In operation, the series and shunt MOSFETs in one arm are turned on and off, respectively, whereas the counterparts in the other arm are respectively turned off and on, hence allowing a RF signal to pass from the input (Port 1) to the output (Port 2 or 3) accordingly.

Consider the SPDT switch in Fig. 3.4a under bias voltage V_{c1} that causes one branch "on" and the other branch "off." The insertion loss (IL) and isolation (ISO) of the series SPDT, or the on- and off-state branches, can be derived from the equivalent circuit in Fig. 3.4b as

$$IL = -20 \log_{10} \left| \frac{2Z_0}{2Z_0 + R_{on\_M1}} \right| \tag{3.7}$$

$$ISO = -20 \log_{10} \left| \frac{2Z_0}{2Z_0 + 1/j\omega C_{off\_M2}} \right| \tag{3.8}$$

where Z_0 is the terminating impedance at ports 1, 2 and 3, and R_{on_M1} and C_{off_M2} are the on-resistance and off-capacitance of transistor M1 and M2, respectively.

Equation (3.7) is essentially the same as (3.1) and shows that the insertion loss of the SPDT switch decreases as R_{on} is reduced, which can occur when the width of M_1 and M_2 is increased. Increasing a transistor's width, however, also increases parasitic capacitance, resulting in more signal leakage from input to output, hence leading to reduced isolation for the switch.

Now, we consider the SPDT switch made of series and shunt MOSFET in each arm as shown in Fig. 3.4c. When V_{c2} is biased to make the SPST branch between Port 1 and 2 "on" and that between Port 1 and 3 "off", the SPDT switch can be modeled as a circuit shown in Fig. 3.4d. The on-state between Port 1 and 2 and the off-state between Port 1 and 3 function as simple RC circuits with their (3-dB) cut-off frequencies obtained as

$$f_{c\_on} = \frac{1}{2\pi R_{on\_Mse1} C_{off\_Msh1}} \tag{3.9}$$

$$f_{c\_off} = \frac{1}{2\pi R_{on\_Msh2} C_{off\_Mse2}} \tag{3.10}$$

respectively, where R_{on_Mse1} (R_{on_Msh2}) is the on-resistance of transistor M_{se1} (M_{sh2}) and C_{off_Msh1} (C_{off_Mse2}) is the off-capacitance of M_{sh1} (M_{s22}), respectively. As can be seen in (3.9) and (3.10), the 3-dB cut-off frequency (f_c), and hence the operating frequency range (DC to f_c) of the series-shunt SPDT can be increased by using transistors having small R_{on} and C_{off}.

The insertion loss and isolation of the series-shunt SPDT switch can be derived from the equivalent circuit in Fig. 3.4d as

$$IL = -20\log_{10}\left|\frac{2Z_0}{2Z_0 + \frac{1}{1/R_{on\_Mse1}+j\omega C_{off\_Msh1}}}\right| \tag{3.11}$$

$$ISO = -20\log_{10}\left|\frac{2Z_0}{2Z_0 + \frac{1}{R_{on\_Msh2}+j\omega C_{off\_Mse2}}}\right| \tag{3.12}$$

Equations (3.7)–(3.12) show that series-shunt SPDT switch as shown in Fig. 3.4c can have lower insertion loss and higher isolation than those for the series SPDT switch shown in Fig. 3.4a, assuming same size for the transistors ($M_1 = M_2 = M_{se1} = M_{se2}$). That is, effectively, series transistors (M_{se1} and M_{se2}) and shunt transistors (M_{sh1} and M_{sh2}) can be employed together to enhance the insertion loss and isolation of a switch, respectively.

Similar to the series SPDT switch, increasing the width of the series transistors (M_{se1} and M_{se2}) decreases the insertion loss of the series-shunt SPDT switch, but reducing the isolation due to increased parasitic capacitance. On the other hand, increasing the width of the shunt transistors (M_{sh1} and M_{sh2}) can improve the isolation of the series-shunt SPDT switch, but can also result in more signal leakage from the input to ground. Therefore, trade-off between series and shunt transistors needs to be considered, and an optimum width for each transistor needs to be determined for optimum performance of the switch.

Figure 3.5 shows the insertion loss and isolation responses of a series-shunt SPDT switch using some nMOSFETs. It can be seen that the insertion loss increases and the isolation decreases as the frequency increases in a low-pass and high-pass trend, respectively, which is undesirable for high-frequency applications. Therefore, the series-shunt SPDT switch is more suitable in the microwave range (to 30 GHz) and should not be employed for applications at millimeter-wave frequencies, unless proper transistors and/or design modifications are implemented.

T/R switch is essentially a SPDT switch. T/R switch is typically used between an antenna, transmitter and receiver to facilitate the transmission and reception of RF signals from the transmitter and receiver, respectively, using a single antenna. The input port of a T/R switch is connected to an antenna, typically referred to as Antenna

Fig. 3.5 Simulated insertion loss and isolation between Port 1–2 and 1–3, respectively, for the series-shunt SPDT switch in Fig. 3.4c

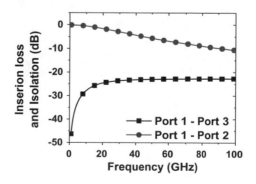

(ANT) port, while the two output ports are connected to a transmitter and receiver, normally referred to as transmitter (TX) and receiver (RX) port, respectively.

3.5 Summary

This chapter covers the fundamentals of switches including insertion loss, isolation, power handling and nonlinearity, and figure of merit. Additionally, basic SPST and series, series-shunt SPDT and T/R switches, forming the basis of switch designs, are also addressed.

Reference

1. Nguyen C (2015) Radio-frequency integrated-circuit engineering. Wiley, New York

Chapter 4
Switching MOSFET

4.1 Introduction

The switching in switches employing MOSFETs is provided by MOSFETs, making these transistors the most important element. Accordingly, the fundamentals of MOSFETs, especially its models and operations in switching, are vital for understanding and designing switches. Advances in CMOS technology have pushed the operation of MOSFETs for switches throughout the microwave region (1–30 GHz) and into the lower region of the millimeter-wave spectrum (30–300 GHz).

This chapter presents the fundamentals of switching MOSFETs useful for the design of switches including models and operations of MOSFETs. Particular MOSFETs realized using deep n-well for enhanced switching performance are also included.

4.2 Models and Operations of MOSFET in Switching

For switches or, in general, RF components involving non-amplification functions such as phase shifters, the MOSFETs employed in these circuits are operated as passive devices. Hence, in these circuits, "passive MOSFET models" are needed. Passive models can be obtained directly using the MOSFET models provided by a foundry with proper bias conditions. They can also be developed by RFIC designers, particularly when the desired operating frequencies are beyond those of available foundry models.

Passive MOSFETs are operated under two conditions: *on* and *off*. When the gate-source voltage V_{gs} is higher than the threshold voltage V_t, the transistor is (theoretically) turned *on*, under which a small resistance R_{on} appears between the drain and source , assuming negligible parasitics. V_{gs} is normally set to V_{DD} (the voltage

© The Author(s) 2020
C. Nguyen and Y. Um, *Multiband Dual-Function CMOS RFIC Filter-Switches*,
SpringerBriefs in Electrical and Computer Engineering,
https://doi.org/10.1007/978-3-030-46248-2_4

supplied to the drain) in practical circuits for on-state. R_{on} is basically the (output) drain-source resistance R_{ds}, which approximately constitutes a simple "on-model" between the drain and source as shown in Fig. 4.1a. On the other hand, when V_{gs} is lower than V_t, the transistor is (theoretically) switched *off*. Typically, V_{gs} is set to 0 V in practical circuits for off-state. It should be noted that the off-state only occurs well at low frequencies. At high frequencies, the parasitic drain-source capacitance, C_{ds}, provides a path for signals from drain to source, hence degrading the off-state condition of the device. A simple off-model between the drain and source can be approximately represented by equivalent-circuit models shown in Fig. 4.1b, c, which are dominated by C_{ds}. The off-capacitance C_{off} and off-resistance R_{off} in Fig. 4.1c can be derived from Fig. 4.1b as

$$C_{off} \simeq C_{ds} + \frac{1}{2}C_{gs} \qquad (4.1)$$

assuming $C_{gs} \simeq C_{gd}$ and

Fig. 4.1 Simple small-signal *on* (**a**) and *off* (**b, c**) models for MOSFETs operating in passive mode. R_{on}, C_{off}, and R_{off} should be as small and large as possible, respectively. Reprinted, with permission, from [1]

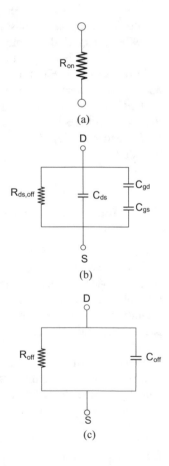

Fig. 4.2 An equivalent-circuit for the on- and off-model of passive MOSFETs. Reprinted, with permission, from [1]

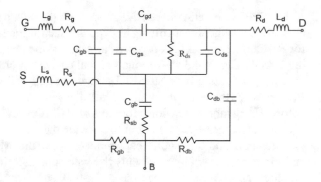

$$R_{off} \simeq R_{ds,off} \tag{4.2}$$

where $C_{ds,off}$ is C_{ds} under off-state condition. Passive MOSFETs are operated between the drain (input or output) and source (output or input) with a control voltage applied to the gate terminal through a typically large resistor, thus acting essentially as a two-terminal device. There is no dc power consumption in passive MOSFETs since there is no current flow at the gate.

More accurate on- and off-model for passive MOSFETs can be described in an equivalent circuit as shown in Fig. 4.2, which represents a four-terminal MOSFET. Typically, the source (S) and bulk (B) terminals are tied together.

L_g, L_d, and L_s represent the gate, drain, and source electrode inductance. R_g, R_d, and R_s are the gate, drain, and source terminal resistance. These resistances are essentially due to the conductivity of the terminal material and the contact. R_{gb}, R_{db}, and R_{sb} are the substrate resistances between the gate and bulk, drain and bulk, and source and bulk, respectively, and are the main sources of loss in MOSFET, primarily due to the lossy Si substrate. R_{ds} represents the drain-source resistance extracted from the drain-source current. C_{ds} is the drain-source capacitance, accounting for the coupling between drain and source due to the lossy Si substrate.

C_{gs} and C_{gd} represents the total capacitance between the gate and source and gate to drain, respectively. These capacitances are basically the overlapping capacitances between the gate and source and gate and drain, including fringing capacitances, resulting from diffusion. The fringing capacitances occur along all the edges of the electrodes and only contribute lightly to the overall capacitances. C_{gb} represents the total capacitance between the gate and bulk (substrate). C_{db} is the junction capacitance between the drain and bulk.

The foregoing on- and off-model can be obtained by fitting the S-parameters of actual devices measured under on and off bias conditions, respectively, over interested frequencies to the S-parameters calculated from the equivalent-circuit model. The S-parameters for passive MOSFETs operating under on and off conditions can also be obtained directly from a large-signal model by setting proper values for V_{gs}— for example, 1.8 V and 0 V for on and off, respectively, from which on and off equivalent-circuit models can be extracted.

MOSFETs operating in passive mode with the bulk open or floated is especially attractive for switching involving high power, such as transmit-receive (T/R) switches for RF transceivers, as bulk floating can improve the linearity and power handling of the circuits. Figure 4.3 shows small-signal equivalent-circuit models for the on and off states of floating-bulk MOSFETs when the gate is floated using a large resistor. R_{ds} represents the resistive loss in the bulk between the source and drain as mentioned before. Using large gate widths for advanced CMOS devices can produce R_{ds} within several ohms, thereby resulting in low loss in the bulk. C_{gd} and C_{gs} again represent the gate-drain and gate-source capacitances due to the overlapping between the gate and diffusion areas. C_{gb} represents the gate-bulk capacitance. C_{db} and C_{sb} are the junction capacitances between the drain-bulk and source-bulk, respectively. C_{ds} is the drain-source capacitance. These capacitances are described earlier. C_{rb1} and C_{rb2} in Fig. 4.3a represent the distributed capacitances between the inversion layer and bulk. All these parasitic capacitances are on the order of tens of fF for advanced CMOS devices and increase with the device's gate width.

Figure 4.4a, b show simplified small-signal equivalent-circuit on- and off-model of MOSFETs obtained from Fig. 4.3a, b, respectively. The on-model consists of the on-resistance R_{on} in parallel with the on-capacitance C_{on}, which represents the total capacitance C_g seen at the gate consisting of C_{gb}, C_{gd} and C_{gs} in Fig. 4.3a, the on-state bulk capacitance C_{b-on} consisting of C_{db}, C_{sb}, C_{rb1} and C_{rb2} in Fig. 4.3a, and the drain-source resistance R_{ds}. The off-model is represented by the off-capacitor C_{off},

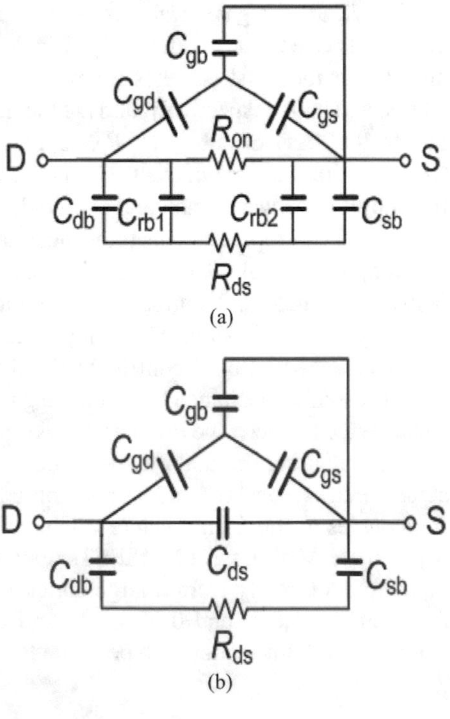

Fig. 4.3 Small-signal equivalent-circuit models for MOSFETs with both the bulk and gate floated under on-state (**a**) and off-state (**b**). Reprinted, with permission, from [1]

Fig. 4.4 Simplified
small-signal
equivalent-circuit models of
the bulk-floated MOSFETs
under on (**a**) and off
(**b**) conditions. Reprinted,
with permission, from [1]

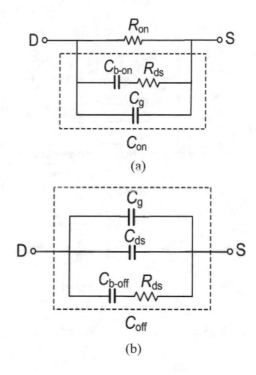

which consists of the total capacitance C_g seen at the gate consisting of C_{gb}, C_{gd} and C_{gs} in Fig. 4.3b, the off-state bulk capacitance $C_{b\text{-}off}$ consisting of C_{db} and C_{sb} in Fig. 4.3b, the drain-source capacitance C_{ds}, and the drain-source resistance R_{ds}.

4.3 Deep n-Well MOSFET for Switching

The main limitation of the CMOS transistors when used in switches is due to their parasitic capacitances, C_{gd} (gate-drain capacitance), C_{gs} (gate-source capacitance), C_{sb} (source-bulk capacitance), and C_{db} (drain-bulk capacitance). The effects of C_{sb} and C_{db} are more significant than those of C_{gs} and C_{gd} at millimeter-wave frequencies due to their larger values and the conductive Si substrate. RF signals can penetrate into the lossy Si substrate via C_{sb} and C_{db}, thus leading to increased insertion loss. To reduce the effects of these capacitances, the floating-body technique with deep n-well terminal of the deep n-well transistors employed in [2] can be used. The bulk are triple-well devices and is floated by a large resistor. With deep n-well, large resistors can be applied directly to the bulk of nMOS devices, making it floated at high frequencies without latch-up problems that would result in circuits consisting of both nMOS and pMOS without deep n-well. Using deep n-well thus allows a switch to be fully integrated with other circuits designed using both nMOS and pMOS

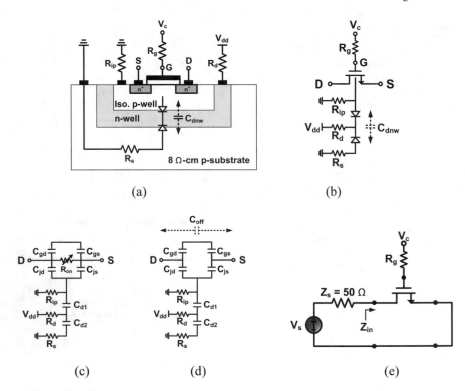

Fig. 4.5 **a** Cross sectional view, **b** schematic, **c**, equivalent-circuit model under different bias conditions, and **d** equivalent-circuit model under off-state (when $V_c = 0$ V) of deep n-well transistors. **e** Simulation set-up for calculating R_{on} and C_{off}. Reprinted, with permission, from [1]

transistors in a single chip. Floating the bulk forces the bulk resistance underneath the source and drain junctions open with respect to the ground, leading to a much smaller resistive loss in the conductive p-bulk than with the bulk grounded.

Figure 4.5a, b show a cross sectional view and schematic of nMOS transistors with deep n-well. R_s represents the losses of the conductive substrate and substrate-ground connection, which are normally small for good substrate and substrate-ground connection. Since the Si substrate is lossy, the effects of the substrate should be included in a transistor model and depend strongly on the layout of circuits. The deep n-well structure is used to isolate the bulk of the transistor from the p-substrate which is connected to the ground, thereby preventing the leakage (and hence loss) of RF signals from the transistor's conducting channel to the ground. However, the deep n-well layer also inadvertently creates a pair of parasitic junction diodes from the junctions between n-well/p-well and n-well/p-substrate, which are effectively equivalent to a parasitic capacitance C_{dnw}.

To maximize the isolation to the substrate provided by the deep n-well layer (and hence minimizing the loss of RF signals), the isolated bulk (p-well) and deep n-well of the transistor are biased at 0 V and 1.8 V (V_{dd}) with large resistors ($R_{ip} = 10$ kΩ

and $R_d = 20\ k\Omega$), respectively, to make the transistor body floating and to establish a reverse-bias for the two p-n deep n-well junctions for reduced C_{dnw}.

Figure 4.5c shows the equivalent-circuit model of deep n-well nMOS transistors under different bias operations and Fig. 4.5d shows the model under off-state. For simplicity in evaluating the on- and off-state models and their corresponding quality (Q) factors, we can assume that when the transistor is biased in on-state, it is approximately equivalent to the on-state resistor (R_{on}) in parallel with a capacitor C_{eq} under on-state, while under off-state, it behaves approximately as C_{eq} under off-state (or equivalently, the off-state capacitor C_{off}). C_{eq} consists of C_{gs} (gate-source), C_{gd} (gate-drain), C_{js} and C_{jd} (junction), and C_{dnw} (deep n-well) comprising C_{d1} and C_{d2} capacitors shown in Fig. 4.5c, d [3].

R_{on} and C_{off} can be determined using the simulation set-up shown in Fig. 4.5e according to

$$R_{on} = real(Z_{in\_on})$$

$$C_{off} = \frac{1}{2\pi f * imag(Z_{in\_off})} \qquad (4.3)$$

where Z_{in_on} and Z_{in_off} are the simulated input impedances when the deep n-well nMOS transistor is biased in on- and off-state, respectively.

It is noted that the Q-factor of the off-state transistor, and hence that of its equivalent off-capacitor C_{off}, affects the performance of switches employing such transistors. Figure 4.6 shows the simulated Q-factor of two transistors (M_{24} and M_{60}) with and without deep n-well biased in off-state using the simulation setup in Fig. 4.5e obtained as

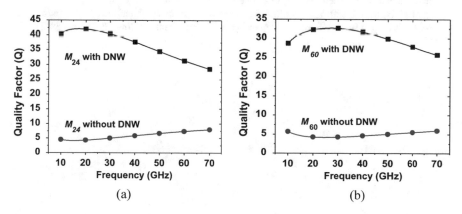

(a) (b)

Fig. 4.6 Simulated Q for off-state nMOS transistors with and without deep n-well (DNW): **a** M_{24} and **b** M_{60}

$$Q = \frac{1}{2\pi f * real(Z_{in\_off}) * C_{off}}$$ (4.4)

M_{24} and M_{60} have width/length of 420 μm/0.18 μm and 240 μm/0.18 μm, respectively. The simulated Q-factors for transistors with/without deep n-well are 41.5/4.5 and 27.8/5.5 at 24 and 60 GHz, respectively. As shown in Fig. 4.6, it is verified that the Q-factor of the off-state transistors with deep n-well is much higher than that of the off-state transistors without deep n-well, making nMOS transistors with deep n-well attractive for achieving low insertion loss in switch design.

4.4 Summary

This chapter covers the fundamentals of switching MOSFETs useful for the design of various switches including general and on/off models as well as operations of MOSFETs. Particular MOSFETs realized using deep n-well for enhanced switching performance are also included. The Q-factor of off-state transistors affecting switch performance, which is often overlooked, is also addressed.

References

1. Nguyen C (2015) Radio-frequency integrated-circuit engineering. Wiley, New York
2. Yeh M-C, Tsai Z-M, Liu R-C, Lin K-Y, Chang Y-T, Wang H (2006) Design and analysis for a miniature CMOS SPDT switch using body-floating technique to improve power performance. IEEE Trans Microw Theor Tech 54(1):31–39
3. Huynh C, Nguyen C (2011) New ultra-high-isolation rf switch architecture and its use for a 10–38-GHz 0.18-μm BiCMOS ultra-wideband switch. IEEE Trans Microw Theor Tech 59(2):345–353

Chapter 5
Design of CMOS Dual-Band Dual-Function Filter-Switches

5.1 Introduction

As mentioned in Chap. 1, multi-band and multi-function RF components including RF switches possess numerous advantages as compared to their counterparts operating only over a single band and single function. In typical RF systems, transmitting and receiving signals in transmitters and receivers transmitted and received through a T/R switch, respectively, are shaped in a frequency window to reject unwanted out-of-band signals. This is normally achieved by routing the signals through external band-pass filters. While these separate band-pass filters can be implemented as off- or on-chip components, they increase the size and cost of overall systems. A more effective approach is to integrate the band-pass filter function with switches to make a dual-function (switching and filtering) simultaneously.

In this chapter, we will discuss first a multiband dual-function filter-switch design concept. We will then describe the designs of the following different SPDT and T/R switches operating over dual-bands centered at 24, 40 and 60 GHz, possessing dual-functions of band-pass filtering and switching, on 0.18 μm CMOS:

- A 0.18 μm CMOS dual-band T/R switch having a band-pass filtering function and multiple ports working concurrently over 35.5–43.7 GHz and 56.4–63 GHz.
- A 0.18 μm CMOS concurrent T/R switch having dual-band-pass filtering function across 17.2–27.3 GHz and 52.5–66.5 GHz, implemented with dual-band quarter-wavelength LC networks and dual-band resonators.
- 0.18 μm CMOS dual-band SPDT and T/R switches having band-pass filtering function over 14.6–30.4 GHz, 48–62.3 GHz and 14.6–30.3 GHz, 48.4–64.8 GHz, respectively.
- A 0.18 μm CMOS concurrent dual-band SPDT filter-switch operating over two distinctive wide bands around 24 and 60 GHz.

© The Author(s) 2020
C. Nguyen and Y. Um, *Multiband Dual-Function CMOS RFIC Filter-Switches*,
SpringerBriefs in Electrical and Computer Engineering,
https://doi.org/10.1007/978-3-030-46248-2_5

5.2   Multi-band Dual-Function Filter-Switch Design

As mentioned in Sect. 3.4, the insertion loss and isolation of typical series-shunt switches show low-pass and high-pass responses due to their inherent structural characteristics. Consequently, it is rather difficult to design a switch to have single or multi-band band-pass filtering function based on the inherent characteristics of series-shunt switches.

To achieve both switching and band-pass filtering in RF systems, external band-pass filters and wideband switches are commonly utilized together as depicted in Fig. 5.1a. This conventional scheme employs two individual components, hence leading to a larger circuit size and hence higher cost, besides other potential issues such as repeatability, reliability, integration, matching, and loss. A more effective approach is to integrate the band-pass filtering function into switches to make a dual-function (switching and filtering) simultaneously as shown in Fig. 5.1b.

To integrate the band-pass function into a switch, we utilize the impedance- or admittance-transformation characteristic of a quarter wavelength together with a transistor-based resonator as shown in Fig. 5.2. Figure 5.2a shows a quarter-wavelength ($\lambda/4$) transformer connected with parallel resonator and Fig. 5.2b shows the equivalent with capacitor C_r of the resonator replaced with a shunt transistor M_{sh} controlled by a biasing voltage V_{ctrl} at the gate. The width of M_{sh} is determined by its off-state capacitance, which is the same as C_r, considering trade-off between insertion loss and isolation and operating bandwidth of the resulting switch.

Figure 5.3b, d show the equivalent circuits of the circuit in Fig. 5.2b under two different bias conditions: $V_{ctrl} = 1.8$ and 0 V according to Fig. 5.3a, c, respectively.

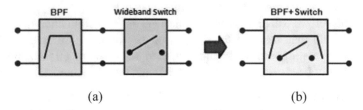

(a)   (b)

Fig. 5.1 Band-pass filtering switch design concept: **a** combination of individual band-pass filter (BPF) and switch and **b** integration of band-pass and switching function in a single circuit

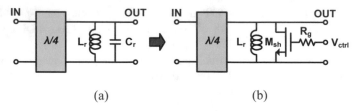

(a)   (b)

Fig. 5.2 a Quarter-wavelength transformer and parallel resonator and **b** quarter-wavelength transformer and parallel resonator simulated with inductor and nMOS transistor

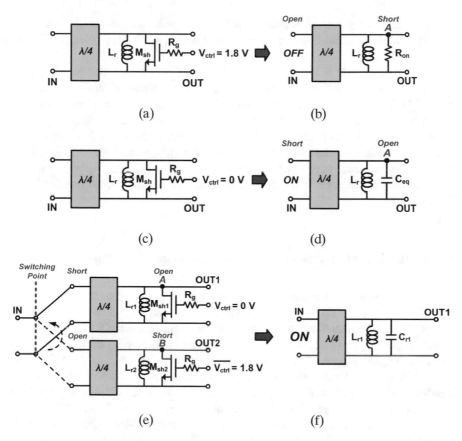

Fig. 5.3 Basis and operation of the proposed band-pass filtering switches: Band-pass filtering switch when $V_{ctrl} = 1.8$ V (**a**) and its equivalent circuit (**b**), band-pass filtering switch when $V_{ctrl} = 0$ V (**c**) and its equivalent circuit (**d**), and SPDT filter-switch (**e**) and its equivalent circuit between IN and OUT1 (**f**)

For simplicity, we may assume R_{on} and C_{eq} under the respective on- and off-state are negligibly small as in ideal cases. Accordingly, we may assume node A in the circuits of Fig. 5.3b, d is short and open, respectively. Consequently, the input port of these circuits becomes open and short, signifying the OFF and ON states of the circuit, as denoted in Fig. 5.3b, d, respectively. Moreover, these circuits also show a band-pass characteristic due to their parallel resonators. It is thus confirmed that the shunt nMOS in these circuits can function not only as a switching, but also as a band-pass circuit, which forms the basis for our filter-switch designs.

The foregoing filter-switch design concept can now be applied to the design of a SPDT filter-switch as shown in Fig. 5.3e, f. This SPDT filter-switch consists of two identical SPST switches as shown in Fig. 5.3e. This SPDT switch is equivalent to a SPST switch according to the control voltages (V_{ctrl}) as shown in Fig. 5.3f. The common node of the two constituent SPST switches at the IN port can be defined as the "switching point" as seen in Fig. 5.3e. This filter-switch design concept is also applicable to multi-band band-pass filtering and multi-port switch as will be implemented later.

A $\lambda/4$ transformer is typically realized with a transmission line in microwave circuits. Transmission-line $\lambda/4$ transformers, however, are not desirable for silicon-based RFICs due to their relatively large size even at millimeter-wave frequencies. To alleviate this issue, we will implement admittance inverters (J-inverter) for parallel resonators normally utilized in band-pass filters, recognizing that J-inverters have $\lambda/4$ characteristics similar to $\lambda/4$ transformers. The design of single- and dual-band switches having band-pass filtering characteristics presented in this book hence is based on not only the switch design concept described previously, but also band-pass filter theory.

5.3 Multi-port Dual-Band Band-Pass Filter-Switches

This section describes a band-pass filter, SPDT band-pass filter-switch, and T/R band-pass filter switch, realized with 0.18 μm CMOS in a 0.18 μm SiGe BiCMOS process [1], operating simultaneously in two different frequency bands of 35.5–43.7 GHz in Q-band (33–50 GHz) and 56.5–63 GHz in V-band (40–75 GHz) centered about 40 and 60 GHz, respectively. The architecture, design and operation of the 40/60 GHz dual-band T/R band-pass filter-switch are based on those of 40/60 GHz dual-band band-pass filter and SPDT band-pass filter-switch. Therefore, the band-pass filter and SPDT filter-switch are described first followed by the T/R filter-switch.

The T/R switch has multiple ports with concurrent dual-band characteristics at each port and high isolation between them, enabling its versatile implementation in multi-band RF systems. The operating frequencies can be used in a multi-band RF system for possible long range in Q-band and short range in V-band operations at the same time, considering the relatively low and high atmospheric attenuations at these respective frequencies, hence extending the application ranges for multi-band RF systems at millimeter-wave frequencies.

5.3.1 40/60 GHz Dual-Band Band-Pass Filter

The design of the 40/60 GHz dual-band SPDT band-pass filter-switch [2] is based on a dual-band band-pass filter using admittance (J) inverters and dual-band resonators described in Sect. 2.7. Figure 5.4 shows the evolution of the dual-band resonator, starting from the conventional dual-band resonator in Fig. 5.4a, which is Fig. 2.13 in Sect. 2.7, to the modified dual-band resonator in Fig. 5.4c.

The conventional dual-band resonator is designed with $C_2 = 26\,f$F, $L_2 = 450$ pH, $C_1 = 250\,f$F, and $L_r = 45\,p$H calculated from (2.33) to (2.36). This resonator is

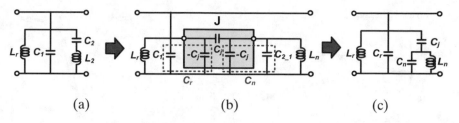

Fig. 5.4 Evolution of the dual-band resonator: **a** conventional dual-band resonator, **b** modified dual-band resonator implementing J-inverter, and **c** final modified dual-band resonator. Reprinted, with permission, from [2]

transformed into another resonator in Fig. 5.4b using a J-inverter consisting of a pi-network of capacitors as shown in Fig. 2.12b of Sect. 2.7. This circuit transformation is needed to obtain L_n of lower inductance and higher Q-factor than that of L_2. Using such a transformed resonator with smaller L_n helps ease the layout and reduce the inductor's physical size. Finally, the modified dual-band resonator in Fig. 5.4c can be obtained as shown. The series L_2–C_2 in Fig. 5.4a is equivalent to the combined J-inverter and parallel C_{2_1}–L_n in Fig. 5.4b, leading to [3]

$$C_n = C_{2\_1} - C_j = L_2 J^2 - \frac{J}{\omega_0} \tag{5.1}$$

$$C_r = C_1 - C_j = C_1 - \frac{J}{\omega_0} \tag{5.2}$$

$$L_n = \frac{C_2}{J^2} \tag{5.3}$$

where $J = 1/Z_0$ and ω_0 is the center frequency of the two pass-band frequencies. From (5.1) to (5.3), C_n, C_r, L_n and C_j of the modified dual-band resonator can be calculated. It is noted that L_n of the modified dual-band resonator is smaller than L_2 of the conventional dual-band resonator, facilitating its design for higher Q, which leads to possibly lower insertion loss for the dual-band resonator, as shown in Fig. 5.5a, b, respectively.

Figure 5.6a, b show the schematic of the dual-band band-pass filter employing the dual-band resonator shown in Fig. 5.4c and its simulated return loss and insertion loss. The dual-band band-pass filter is realized by the third-order Chebychev approximation with 0.01 dB ripple, which is chosen for optimal results considering trade-off between insertion loss, out-of-band rejection ratio and isolation of the switch, and 20% fractional bandwidth at both center frequencies of 40 and 60 GHz. The corresponding element values are listed in Fig. 5.6a, where the source Z_S and load Z_L impedances are 50 Ω. The identical J-inverters ($J_{01} = J_{34}$) and ($J_{12} = J_{23}$) are implemented using the pi-networks of capacitors and inductors as shown in Fig. 2.12b, c, respectively. To facilitate the implementation of J_{01} and J_{34}, these inverters are replaced by equivalent L-type networks of capacitors, which are derived from the

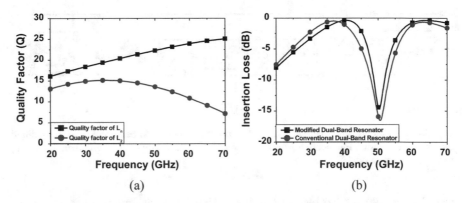

Fig. 5.5 Comparison of simulated quality (Q) factor of inductors L_n and L_2, and insertion losses of modified and conventional dual-band resonators: **a** Q-factors and **b** insertion losses. Reprinted, with permission, from [2]

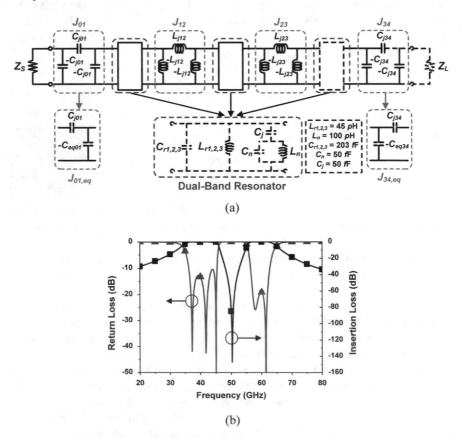

Fig. 5.6 **a** Schematic of the 3rd order dual-band band-pass filter using dual-band resonators and J-inverters and **b** its return loss and insertion loss. Reprinted, with permission, from [2]

pi-type network of capacitors in Fig. 2.12b, as shown in Fig. 5.6a. C_{eq01} and C_{eq34} are combined with C_{r1} and C_{r3}, respectively, and the total capacitances C_{T1} and C_{T3} are obtained by

$$C_{eq01} = C_{eq34} = \frac{C_{j01}}{1 + (\omega_0 C_{j01}/Y_0)^2} = \frac{C_{j34}}{1 + (\omega_0 C_{j34}/Y_0)^2} \tag{5.4}$$

$$C_{T1} = C_{T3} = -C_{eq01} + C_{r1} = -C_{eq34} + C_{r3} \tag{5.5}$$

where Y_0 is the reference admittance. Due to the high-pass (J_{01} and J_{34}) and low-pass (J_{12} and J_{23}) responses of the employed J-inverters, the filter can achieve similar degree of out-of-band rejection ratio in both lower (40 GHz) and upper (60 GHz) frequency bands.

5.3.2 40/60 GHz Dual-Band SPDT Band-Pass Filter-Switch

Figure 5.7 shows the 40/60 GHz SPDT switch with integrated dual-band band-pass filtering function, in which the signal path from the IN port to the OUT1 (or OUT2) port is designed based on the designed dual-band band-pass filter described in Sect. 5.3.1. The shunt nMOS transistors M_1, M_2, and M_3 in Fig. 5.7 not only provide the switching function, but also simulate the capacitors C_{T1}, C_{r2}, C_{T3} described in Sect. 5.3.1, respectively, constituting the band-pass filtering function. Body-floating technique with deep n-well is employed for all the nMOS transistors to decrease the parasitic capacitances of the transistors [4]. Point A on the figure represents the switching point, at which the SPDT switch turns ON for one output port (e.g., from IN to OUT1) and OFF for another port (e.g., from IN to OUT2). Detailed switching operation is explained as follows.

The switching function is executed through J_{12}, J_{23}, M_2 and M_3. For instance, when M_1 is biased at $V_{C1} = 0$ V (off-state) while M_2 and M_3 on the OUT1 and OUT2 paths are biased at $V_{C2} = 0$ V (off-state) and $\overline{V_{C2}} = 1.8$ V (on-state), the

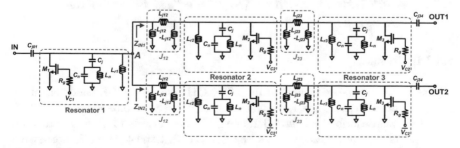

Fig. 5.7 Schematic of the 40/60 GHz dual-band SPDT band-pass filter-switch. Reprinted, with permission, from [2]

OUT1 port is approximately matched to the IN port via the switching point A, while A appears as an approximate open circuit looking toward the OUT2 port through the J-inverters and on-state shunt transistors. Note that the J-inverter behaves as a quarter-wavelength transmission line and hence can transform a low impedance caused by an on-state shunt transistor to a high impedance and a high impedance by an off-state shunt transistor to a low impedance at A in Fig. 5.7. When $V_{C2} = 0$ V in Fig. 5.7, the path from IN to OUT1 is in on-state with Z_{IN1} at A to be 41 and 56 Ω simulated at 40 and 60 GHz, respectively. On the other hand, OUT2 becomes isolated from OUT1, and the simulated impedances Z_{IN2} looking into OUT2 at A are 236 and 538 Ω at 40 and 60 GHz, respectively. Therefore, the SPDT switch in Fig. 5.7 is equivalent to a SPST switch (upper path), which is equivalent to the dual-band band-pass filter with off-state shunt transistor as shown in Fig. 5.6a, while the lower path at the switching point A is in off-state.

It should be noted that the conceived SPDT filter-switch architecture facilitates extensions to dual-band band-pass filtering single-pole multi-throw switches by adding extra output paths, while maintaining desired filtering and switching functions with simplicity and compactness.

The isolation between the output ports OUT1 and OUT2 primarily depends on the impedance looking into the OFF-output port at the switching point, which is proportional to the numbers of J-inverters and shunt transistors in the OFF output path. The considered SPDT filter-switch has one common resonator (Resonator 1) for both OUT1 and OUT2 paths with switching right before J_{12}, hence producing higher isolation than that using two common resonators (Resonator 1 and 2) with switching before J_{23}. This design concept is useful as it enables high isolation with less number of sections. This is particularly attractive since with conventional design, using less resonators would improve the insertion loss, yet decreasing the isolation and possibly degrading the switch's performance. For instance, removing the resonators right before OUT1 and OUT2 results in not only a reduction of the isolation, but also a deterioration of the 3rd dual-band band-pass filter performance including out-of-band rejection ratio, and hence that of the SPDT filter-switch.

5.3.3 40/60 GHz Dual-Band T/R Band-Pass Filter-Switch (Design 1)

5.3.3.1 Design

Figure 5.8 shows the schematic of the 40/60 GHz dual-band T/R band-pass filter-switch, referred to as Design 1, which is realized using a combination of three SPDT filter-switches described in Sect. 5.3.2: one SPDT between Port 1, 2 and 3; one SPDT between Port 3, 1 and 5; and another between Port 2, 1 and 4. The T/R filter-switch consists of five ports that work in the 40 and 60 GHz bands simultaneously: Port 1 (TX) is the transmitting port; Port 2 (ANT1) and Port 3 (ANT2) are the antenna

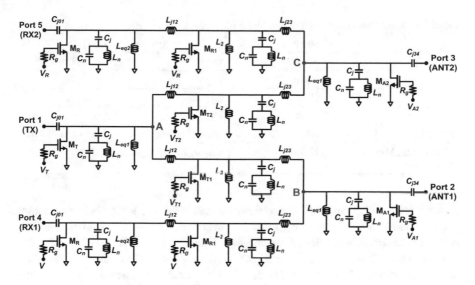

Fig. 5.8 Schematic of the 40/60 GHz dual-band T/R band-pass filter-switch. Reprinted, with permission, from [2]

ports; and Port 4 (RX1) and Port 5 (RX2) are the receiving ports. The equivalent shunt inductances L_{eq1}, L_{eq2}, and L_2 can be determined from

$$L_{eq1} = \left(\frac{1}{L_{r1}} + \frac{1}{-L_{j12}} + \frac{1}{-L_{j12}} \right)^{-1} = \left(\frac{1}{L_{r3}} + \frac{1}{-L_{j23}} + \frac{1}{-L_{j23}} \right)^{-1} \quad (5.6)$$

$$L_{eq2} = \left(\frac{1}{L_{r1}} + \frac{1}{-L_{j12}} \right)^{-1} \quad (5.7)$$

$$L_2 = \left(\frac{1}{L_{r2}} + \frac{1}{-L_{j12}} + \frac{1}{-L_{j23}} \right)^{-1} \quad (5.8)$$

where L_{r1}, L_{r2} and L_{r3} are the inductors of the 1st, 2nd and 3rd dual-band resonators, respectively, and $-L_{j12}$ and $-L_{j23}$ are the negative inductors of J_{12} and J_{23}, respectively. Part of L_{eq1} and L_{eq2} is used for J_{12} and J_{23} along with the series elements L_{j12} and L_{j23}, and part is used for the 1st and 3rd dual-band resonators, respectively. Also, part of the capacitor representing the transistor right after C_{j01} and right before C_{j34} is used for J_{01} and J_{34}, respectively. All the element values for the final design are listed in Table 5.1.

The T/R filter-switch operates in two different operation modes, transmission and reception, each can be inferred from that of the constituent SPDT filter-switches. The receiving operations between ANT1–RX1 and ANT2–RX2 occur simultaneously and happen when M_R, M_{R1}, M_{A1} and M_{A2} of the receiving paths between ANT1–RX1 and ANT2–RX2 are turned off and M_T, M_{T1} and M_{T2} of the transmitting paths between TX–ANT1 and TX–ANT2 are turned on. The 40/60 GHz signals

Table 5.1 40/60 GHz dual-band T/R band-pass filter-switch's parameters

C_j	$50\,fF$	C_n	$50\,fF$
C_{j01}, C_{j34}	$540\,fF$	L_2	$60\,pH$
L_n	$100\,pH$	L_{eq1}	$180\,pH$
L_{eq2}	$90\,pH$	L_{j12}, L_{j23}	$176\,pH$
R_g	$1\,k\Omega$	V_T, V_R, V_{A1}, V_{A2}	0, 1.8 V
M_T, M_{A1}, M_{A2}, M_R	$0.18\,\mu m/208\,\mu m$	M_{T1}, M_{T2}, M_{R1}	$0.18\,\mu m/312\,\mu m$
C_{off} of M_T, M_{A1}, M_{A2}, M_R	$172\,fF$	C_{off} of M_{T1}, M_{T2}, M_{R1}	$253\,fF$

coming from ANT1 and ANT2 are directed into RX1 and RX2 via the switching points B and C, respectively. The transmission between TX–ANT1 (or between TX–ANT2) is obtained when M_T, M_{T1} and M_{A1} between TX–ANT1 (or M_T, M_{T2} and M_{A2} between TX–ANT2) are turned off, while M_{T2}, M_{A2} between TX–ANT2 (or M_{T1}, M_{A1} between TX–ANT1), M_R and M_{R1} are turned on. The 40/60 GHz signals from TX are directed into ANT1 or ANT2 via the switching points A, B or A, C, respectively.

The relations of the isolations between different ports can be deduced from the symmetry of the T/R switch as shown in Fig. 5.8. For instance, the isolations from TX–RX1 (TX–RX2) under transmission from TX–ANT1 (TX–ANT2) and during receptions from ANT1–RX1 and ANT2–RX2 should be similar because of the same numbers of the J-inverters and shunt transistors in the off-state path. Similarly, it is expected that the isolation between ANT1–ANT2 is comparable during transmission from TX–ANT1 or TX–ANT2 and is equal to that of TX–RX1 or TX–RX2.

The 40/60 GHz dual-band T/R band-pass filter-switch was designed and fabricated on a TowerJazz 0.18 μm SiGe BiCMOS process [1]. Figure 5.9 shows a microphotograph of the fabricated T/R filter-switch that occupies 1840 μm × 860 μm excluding all the testing dc and RF pads. To facilitate measurements using a 3-port vector network analyzer, making use of the switch's symmetry, RF test pads are placed only on a half-section of the switch as shown in Fig. 5.9 (Port 1, Port 2, Port 4) with Port 3 and 5 terminated with 50 Ω resistors. The fabricated T/R filter-switch allows the

Fig. 5.9 A microphotograph of the 40/60 GHz dual-band T/R band-pass filter-switch. Reprinted, with permission, from [2]

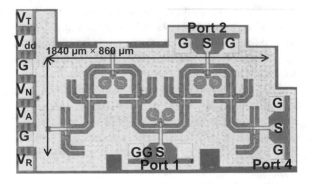

transmitting and receiving operation at 40/60 GHz to be measured between Port 1, Port 2 and Port 4, and the results can be deduced for the operations between other ports.

5.3.3.2 Simulations and Measurements

Figure 5.10 shows the measured and simulated insertion loss, return loss and isolation of the 40/60 GHz T/R filter-switch in the transmitting and receiving modes. The ports corresponding to the measurement parameters are denoted in Fig. 5.9. Figure 5.10a, b show the return loss/insertion loss and isolation results for the reception operation, respectively. The insertion losses (S_{42}) are 8.9 and 12.5 dB at 40 and 60 GHz, respectively. The measured 3 dB bandwidths based on the insertion loss in each passband are 35.1–43.7 and 56.5–63 GHz. The measured input (S_{22}), output

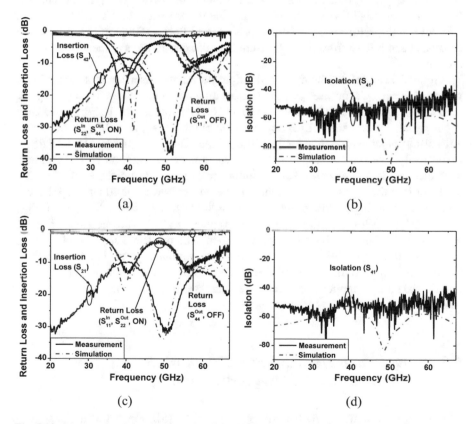

Fig. 5.10 Measured and simulated results of the 40/60 GHz dual-band T/R band-pass filter-switch: **a** return loss and insertion loss, and **b** isolation for receiving operation mode, **c** return loss and insertion loss, and **d** isolation for transmitting operation mode. Reprinted, with permission, from [2]

(S_{44}) return losses are 12, 14 dB at 40 GHz and 9.4, 7.8 dB at 60 GHz, respectively. The drift of the resonance frequency around 40 GHz was due to possible capacitance and inductance variation from the fabrication. The measured isolations (S_{41}) are 56 and 51 dB at 40 and 60 GHz, respectively. The measured stop-band rejection ratio from the lowest insertion loss at 40 GHz to the highest rejection at 52 GHz is 30 dB. Figure 5.10c, d show the return loss/insertion loss and isolation results for the transmission operation, respectively. The insertion losses (S_{21}) are 10 and 12.7 dB at 40 and 60 GHz, respectively. The measured 3 dB bandwidths based on the insertion loss in each passband are 35.5–44.2 and 56.4–63.7 GHz. The 3 dB bandwidths in the transmitting operation are similar to those in the receiving operation. The measured input (S_{11}), output (S_{22}) return losses are 12.7, 12 dB at 40 GHz and 9.4, 8.8 dB at 60 GHz, respectively. The measured isolations (S_{41}) are 57 and 51 dB at 40 and 60 GHz, respectively. The measured stop-band rejection ratio between 40 and 51 GHz is 22 dB. It is verified that the T/R filter-switch has similar TX–RX1 isolation (S_{41}) in the transmission and reception modes, as discussed earlier. The isolation between ANT1–ANT2 should be similar to the measured TX–RX1 isolation. Moreover, the performance at Port 3 (ANT2) and Port 5 (RX2) in Fig. 5.8 can be estimated accurately from these measured results due to the symmetry of the T/R switch.

Figure 5.11a, b show the measured output power and insertion loss versus input power of the switch for three cases in transmitting operation: a 40 GHz input signal, a 60 GHz input signal, and concurrent 40 and 60 GHz input signals. For the 40 GHz input signal, the measured 1 dB compression points (IP_{1dB} and OP_{1dB}) are 23 and 11.6 dBm, respectively, while for the 60 GHz input signal, they are 16.5 and 2.8 dBm, respectively. For the concurrent power measurement, two 40 and 60 GHz signals with identical input power level are simultaneously injected from the vector network analyzer into the input of the switch, and the output powers at 40 and 60 GHz are measured accordingly. With the concurrent 40/60 GHz signals, the IP_{1dB} and OP_{1dB} are 18 and 6.4 dBm at 40 GHz and 14 and 0 dBm at 60 GHz, respectively. For the concurrent 40 and 60 GHz power measurement, the measured IP_{1dB} at 60 GHz is lower than that at 40 GHz because of the higher nonlinear parasitic capacitance of the off-state shunt transistor at 60 GHz. Also, compared to the non-concurrent 40 and 60 GHz single input signals, the P_{1dB} decreases due to 40 and 60 GHz signals' intermodulation when they are injected simultaneously.

5.3.4 Dual-Band T/R Band-Pass Filter-Switch with Dual-Band LC Network (Design 2)

This section presents a T/R band-pass filter-switch [5], referred to as Design 2, operating simultaneously in two different frequency bands of 17.2–27.3 and 52.5–66.5 GHz, which cover the unlicensed bands around 24 and 60 GHz, realized with 0.18 μm CMOS in a 0.18 μm SiGe BiCMOS process [1]. The T/R filter switch

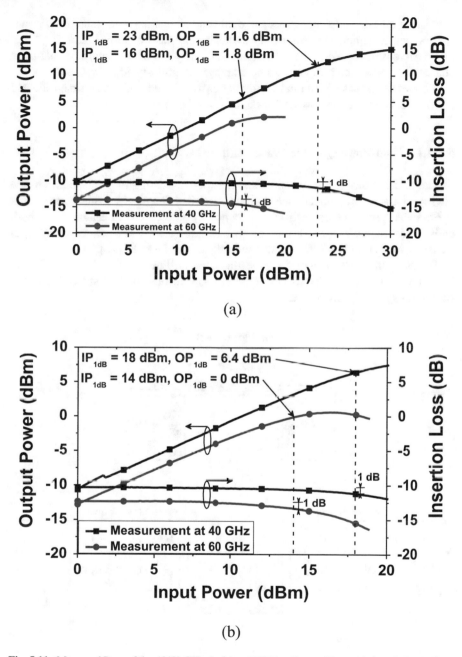

Fig. 5.11 Measured P_{1dB} of the 40/60 GHz dual-band T/R band-pass filter-switch: **a** individual 40 or 60 GHz input signal and **b** concurrent 40 and 60 GHz input signals. Reprinted, with permission, from [2]

consists of dual-band LC networks and dual-band resonators with shunt nMOS transistors, and it demonstrates not only switching but also band-pass filtering functions. Moreover, the T/R filter-switch has multiple ports with concurrent dual-band characteristics at each port and can be employed in multi-band RF systems.

This section starts with a dual-band quarter-wavelength LC circuit first, followed by the T/R band-pass filter-switch designed based on the LC circuit.

5.3.4.1 Dual-Band Quarter-Wavelength LC Circuit

Figure 5.12a shows a transmission-line network consisting of a transmission line having characteristic impedance Z_b and electrical length θ_b loaded with shunt open stubs having characteristic impedance Z_a and electrical length θ_a at both ends. This transmission-line circuit can behave equivalently as a quarter-wavelength transmission line at two different desired frequencies f_1 and f_2 [6].

Equating the ABCD matrix of the transmission-line network in Fig. 5.12a to that of a conventional quarter-wavelength ($\lambda/4$) transmission line at two interested frequencies f_1 and f_2, we obtain

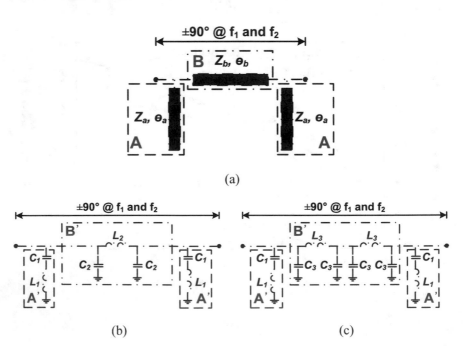

(a)

(b)   (c)

Fig. 5.12 **a** Transmission-line network acting as a quarter-wavelength transmission line at two different frequencies f_1 and f_2 (Model 1), and **b**, **c** its equivalent dual-band LC networks (Model 2 and 3). © [2017] IEEE. Reprinted, with permission, from [5]

$$\begin{bmatrix} 1 & 0 \\ \frac{j\tan\theta_{a\_f1,f2}}{Z_a} & 1 \end{bmatrix} \cdot \begin{bmatrix} \cos\theta_{b\_f1,f2} & jZ_b\sin\theta_{b\_f1,f2} \\ \frac{j\tan\theta_{b\_f1,f2}}{Z_b} & \cos\theta_{b\_f1,f2} \end{bmatrix} \cdot \begin{bmatrix} 1 & 0 \\ \frac{j\tan\theta_{a\_f1,f2}}{Z_a} & 1 \end{bmatrix} = \begin{bmatrix} 0 & \pm jZ_c \\ \pm j\frac{1}{Z_c} & 0 \end{bmatrix}$$

$$(5.9)$$

where Z_c is the characteristic impedance of the conventional $\lambda/4$ transmission line. Relations between Z_a, Z_b, θ_a and θ_b can be obtained from (5.9) as

$$\begin{aligned} Z_{b\_f1} &= \frac{\pm Z_c}{\sin\theta_{b\_f1}} \\ Z_{b\_f2} &= \frac{\pm Z_c}{\sin\theta_{b\_f2}} \end{aligned}$$

$$(5.10)$$

$$\begin{aligned} Z_{a\_f1} &= Z_{b\_f1}\tan\theta_{a\_f1}\tan\theta_{b\_f1} \\ Z_{a\_f2} &= Z_{b\_f2}\tan\theta_{a\_f2}\tan\theta_{b\_f2} \end{aligned}$$

$$(5.11)$$

where the subscripts f_1 and f_2 denote the corresponding frequencies.

Solving (5.10) yields a relation between θ_{b_f1} and θ_{b_f2} as

$$\theta_{b\_f2} = n\pi - \theta_{b\_f1}$$

$$(5.12)$$

where n = 1, 2, 3, ... Utilizing the relation between $\theta_{b\_f1}, \theta_{b\_f2}, f_1$ and f_2 of

$$\frac{\theta_{b\_f1}}{\theta_{b\_f2}} = \frac{f_1}{f_2}$$

$$(5.13)$$

it can then be deduced that

$$\theta_{b\_f1} = \frac{n\pi f_1}{f_1 + f_2} \text{ and } \theta_{b\_f2} = \frac{n\pi f_2}{f_1 + f_2}$$

$$(5.14)$$

When n = 1 corresponding to the shortest length, we get

$$\theta_{b\_f1} = \frac{\pi f_1}{f_1 + f_2} \text{ and } \theta_{b\_f2} = \frac{\pi f_2}{f_1 + f_2}$$

$$(5.15)$$

A solution of (5.11) for θ_{a_f1} and θ_{a_f2} is

$$n\pi \pm \theta_{a\_f1} = \theta_{a\_f2}$$

$$(5.16)$$

where n = 1, 2, 3, ..., and, using the relation between $\theta_{a\_f1}, \theta_{a\_f2}, f_1$ and f_2 of

$$\frac{\theta_{a\_f1}}{\theta_{a\_f2}} = \frac{f_1}{f_2}$$

$$(5.17)$$

we can obtain

$$\theta_{a\_f1} = \frac{n\pi f_1}{f_1 + f_2} \text{ and } \theta_{a\_f2} = \frac{n\pi f_2}{f_1 + f_2}$$

$$(5.18)$$

For n = 1 for the shortest length, we obtain

$$\theta_{a\_f1} = \frac{\pi f_1}{f_1 + f_2} \text{ and } \theta_{a\_f2} = \frac{\pi f_2}{f_1 + f_2} \tag{5.19}$$

Finally, θ_{a_f1}, θ_{a_f2}, θ_{b_f1}, θ_{b_f2}, Z_a and Z_b can be determined from (5.15) to (5.19) and design frequencies f_1 and f_2.

The transmission-line circuit in Fig. 5.12a is not desirable for silicon RFICs due to its large size even at millimeter-wave frequencies. This issue, however, can be overcome by implementing the equivalent lumped-element networks as shown in Fig. 5.12b, c, where the series L–C (A$'$) and the 1st and 2nd order LC pi-networks (B$'$) replace the open stub (A) and cascaded transmission line section (B), respectively.

The required inductance L_1 and capacitance C_1 can be obtained by equating its input impedance to the input impedance of the corresponding open stub as

$$Z_{in}^{sh} = Z_a \frac{Z_L + j Z_a \tan \theta_a}{Z_a + j Z_L \tan \theta_a} = \frac{Z_a}{j \tan \theta_a} \tag{5.20}$$

$$Z_{in}^1 = j\omega L_1 + \frac{1}{j\omega C_1} = \frac{1 - \omega^2 L_1 C_1}{j\omega C_1} \tag{5.21}$$

$$\frac{1 - \omega^2 L_1 C_1}{j\omega C_1} = \frac{Z_{a\_f1}}{j \tan \theta_{a\_f1}} \tag{5.22}$$

from which

$$C_1 = \frac{\tan \theta_{a\_f1}}{\omega Z_{a\_f1}} \left[1 - \left(\frac{\omega}{\omega_0} \right)^2 \right] \tag{5.23}$$

$$L_1 = \frac{1}{C_1 \omega_0^2} = \frac{1}{\tan \theta_{a\_f1}} \cdot \frac{\omega Z_a}{\omega_0^2 - \omega^2} \tag{5.24}$$

where ω_0 is the center frequency between the dual-band center frequencies (f_1 and f_2).

Each half of the 1st order LC pi-network in Fig. 5.12b behave equivalently to a half of the transmission-line circuit in Fig. 5.12a. Hence, we can equate their corresponding ABCD matrices as

$$\begin{bmatrix} \cos \theta_{b\_f1} & j Z_b \sin \theta_{b\_f1} \\ \frac{j \tan \theta_{b\_f1}}{Z_b} & \cos \theta_{b\_f1} \end{bmatrix} = \begin{bmatrix} 1 + \frac{L_2}{C_2} & j\omega L_2 \\ j\omega C_2 (2 - \omega^2 L_2 C_2) & 1 + \frac{L_2}{C_2} \end{bmatrix} \tag{5.25}$$

which yields

$$C_2 = \frac{1 - \cos \theta_{b\_f1}}{\omega Z_b \sin \theta_{b\_f1}} \tag{5.26}$$

$$L_2 = \frac{Z_b \sin \theta_{b\_f1}}{\omega} \tag{5.27}$$

Similarly, each half of the 2nd order LC pi-network in Fig. 5.12c functions equivalently to a half of the transmission-line circuit in Fig. 5.12a, leading to equal ABCD matrices as

$$\begin{bmatrix} \cos \frac{\theta_{b\_f1}}{2} & jZ_b \sin \frac{\theta_{b\_f1}}{2} \\ \frac{j \tan \frac{\theta_{b\_f1}}{2}}{Z_b} & \cos \frac{\theta_{b\_f1}}{2} \end{bmatrix} = \begin{bmatrix} 1 + \frac{L_3}{C_3} & j\omega L_3 \\ j\omega C_3 (2 - \omega^2 L_3 C_3) & 1 + \frac{L_3}{C_3} \end{bmatrix} \tag{5.28}$$

from which,

$$C_3 = \frac{1 - \cos \frac{\theta_{b\_f1}}{2}}{\omega Z_b \sin \frac{\theta_{b\_f1}}{2}} \tag{5.29}$$

$$L_3 = \frac{Z_b \sin \frac{\theta_{b\_f1}}{2}}{\omega} \tag{5.30}$$

Figure 5.13 shows the simulated return losses, insertion losses, phases and input admittances of the transmission-line (Model 1) and lumped-element (Model 2 and 3) networks in a dual-band centered at 24 and 60 GHz. Model 1 and 3 clearly show two pass-bands and one stop-band and λ/4 property (±90°). They also show same input admittance at 24 and 60 GHz when their output ports are terminated with a 50 Ω load. However, even though Model 2 shows λ/4 property (±90°) at 24 and 60 GHz, it does not demonstrate two pass bands clearly and same input admittance at 24 and 60 GHz. As a result, Model 3 is selected as the equivalent LC network of Model 1.

5.3.4.2 24/60 GHz Dual-Band T/R Band-Pass Filter-Switch

Design

Figure 5.14a shows the conventional dual-band resonator (left), mentioned in Fig. 5.4a, and the incorporation of a switching function through replacement of C_r with an shunt nMOS transistor (M_{sh}) (right). The transistor, as mentioned earlier, is approximately equivalent to its on-resistor (R_{on}) and off-capacitor (C_{off}) when 1.8 and 0 V are applied to gate, respectively. Body-floating technique with deep n-well is applied to all the nMOS transistors not only to increase the isolation to the transistors' bulk, but also to decrease the parasitic capacitances associated with the transistors. Figure 5.14b shows the designed T/R band-pass filter-switch with one transmitter (TX) port, two receiver (RX) ports, and two antenna (ANT) ports, where the two identical TX-ANT-RX sections are symmetrically placed with respect to the TX port. All the element values for the final design are listed in Table 5.2. The dual-band resonator with the nMOS transistor shown in Fig. 5.14a is connected in

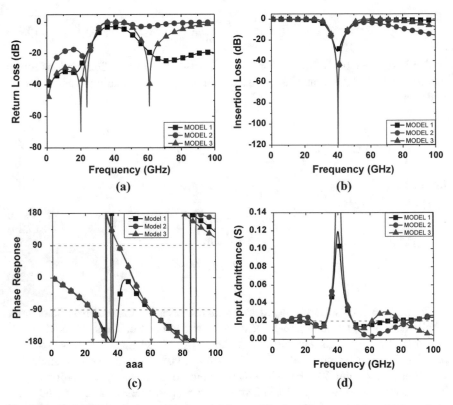

Fig. 5.13 Simulated return losses (**a**), insertion losses (**b**), phase responses (**c**), and input admittances (**d**) of the 24/60 GHz dual-band transmission-line (Model 1) and LC (Model 2 and 3) networks shown in Fig. 5.12

shunt at the end of each port and the dual-band LC network connects ANT ports to the TX and RX ports. In Fig. 5.14b, C_{eq1} and L_{eq1} at A, B and C, and C_{eq2} and L_{eq2} at ports 4 and 5 are obtained as

$$C_{eq1} = C_1 + \frac{C_n}{2} \tag{5.31}$$

$$L_{eq1} = \left(\frac{1}{L_1} + \frac{1}{2L_n} \right)^{-1} \tag{5.32}$$

$$C_{eq2} = C_1 + C_n \tag{5.33}$$

$$L_{eq2} = \left(\frac{1}{L_1} + \frac{1}{L_n} \right)^{-1} \tag{5.34}$$

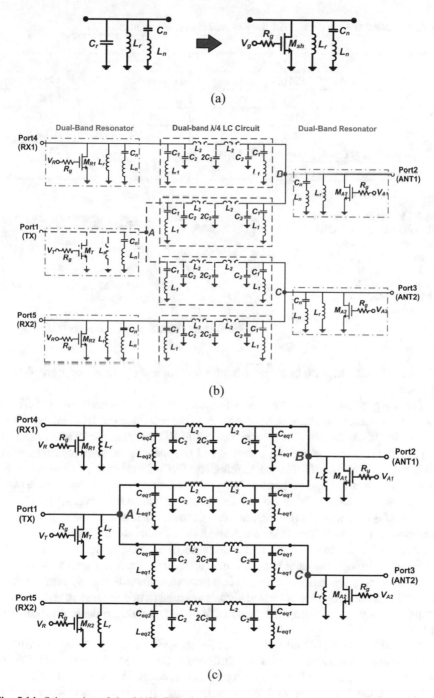

Fig. 5.14 Schematics of the 24/60 GHz dual-band resonator (**a**) and dual-band T/R band-pass filter-switch (**b**, **c**). © [2017] IEEE. Reprinted, with permission, from [5]

Table 5.2 24/60 GHz dual-band T/R band-pass filter-switch's parameters

C_2	20 fF	C_{eq1}	84 fF
C_{eq2}	103 fF	L_2	165 pH
L_r	230 pH	L_{eq1}	193 pH
L_{eq2}	138 pH	Z_a	82 Ω
Z_b	52 Ω	Θ_a, Θ_b	$2\pi/7$
R_g	1 kΩ	V_T, V_R, V_{A1}, V_{A2}	0, 1.8 V
M_T, M_{A1}, M_{A2}, M_R	0.18 μm/102.4 μm	R_{on}	10 Ω
C_{off}	61.5 fF		

Fig. 5.15 Microphotograph of the fabricated 24/60 GHz dual-band T/R band-pass filter-switch. The port numbers correspond to those in Fig. 5.14c. © [2017] IEEE. Reprinted, with permission, from [5]

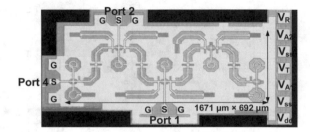

where C_n, L_n are elements of the dual-band resonator and C_1, L_1 are of the dual-band LC circuit.

The designed 24/60 GHz dual-band T/R band-pass filter-switch has two different operation modes: receiving and transmitting. In the receiving mode, as inferred from Fig. 5.14c, M_T at the TX port is turned on while all other transistors are turned off. Under this condition, switching points B and C appear as on-state looking toward the RX1, RX2 ports and off-state looking into the TX port at the two pass-bands' frequencies through the dual-band LC networks and off-state (M_R) and on-state (M_T) transistors, respectively. For the transmitting mode deduced from Fig. 5.14c, M_T and M_{A1} (or M_{A2}) are turned off while the rest of the transistors are turned on. The TX port to ANT 1 (or ANT 2) port via the A and B (or C) is in on-state; on the other hand, the ANT 1 and 2 ports to RX 1 and 2 ports are in off-state at the two pass-bands' frequencies through the dual-band LC network and on-state transistors (M_R). This T/R filter-switch is configured for use in a particular phased-array system in which vertically and horizontally polarized signals are transmitted separately and received simultaneously. It can be implemented in a front-end module similar to that proposed in [7].

The 24/60 GHz dual-band T/R band-pass filter-switch employs CMOS transistors and was fabricated using a TowerJazz 0.18 μm SiGe BiCMOS process [1]. Due to the available RF ports on the vector network analyzer, only three ports at TX, ANT1 and RX1 of the T/R filter-switch could be measured to verify its performance. The un-measured ports (Port 3 and 5 in Fig. 5.14c are terminated with 50 Ω resistors.

Fig. 5.16 Measured and simulated results of the 24/60 GHz dual-band T/R band-pass filter-switch under **a** receiving and **b** transmitting operations. © [2017] IEEE. Reprinted, with permission, from [5]

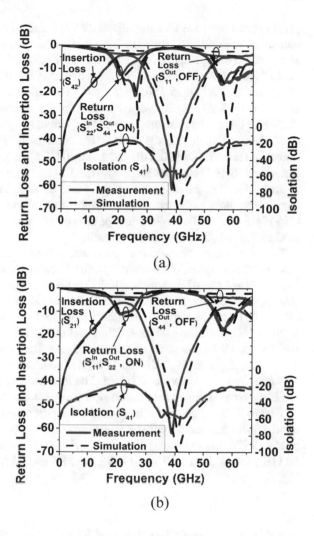

(a)

(b)

Figure 5.15 shows a microphotograph of the fabricated T/R filter-switch that occupies 1671 μm × 692 μm excluding all the dc and RF test pads.

Simulation and Measurement Results

Figure 5.16 shows the measured and simulated insertion losses, return losses, and isolations of the designed T/R filter-switch for the receiving and transmitting operating modes. The measured and simulated results show good agreement. For the receiving mode as shown in Fig. 5.16a, the measured insertion losses (S_{42}) are 4.5 and 5 dB at 24 and 60 GHz, respectively. The measured 3 dB bandwidths are from 17.2 to 27.3 GHz and from 52.5 to 66.5 GHz. The measured input (S_{22}) and output

(S_{44}) return losses are 16 and 8 dB at 24 GHz, and 11.5 and 14 dB at 60 GHz. The measured isolations between the TX (Port 1) and RX (Port 4) are 16 and 18.3 dB at 24 and 60 GHz, respectively. Stop-band rejection of over 40 dB is achieved from 36.2 to 40.8 GHz with the peak rejection of 61.5 dB at 38.4 GHz. For the transmitting mode as shown in Fig. 5.16b, the measured insertion losses (S_{21}) at 24 and 60 GHz are 6.7 and 8.5 dB, respectively. The measured 3 dB bandwidths are 17.2–27.3 and 52.5–66.5 GHz. The measured input (S_{11}), output (S_{22}) return losses are 10, 10.2 dB at 24 GHz and 12.7, 12.6 dB at 60 GHz, respectively. The TX–RX measured isolations (S_{41}) are 18.2 and 20.8 dB at 24 and 60 GHz, respectively. Stop-band rejection exceeding 40 dB is obtained from 35.8 to 41.3 GHz with the peak rejection of 65.5 dB at 38.3 GHz. The insertion losses at 24 and 60 GHz in the transmitting mode are higher than those in the receiving mode because of two switching points (A, B or A, C in Fig. 5.14c) in the transmitting path as compared to only one switching point (B or C in Fig. 5.14c) in the receiving path.

Figure 5.16a, b show the measured output power and insertion loss versus input power of the designed T/R filter-switch for different transmitting modes. When single tones at 24 and 60 GHz are applied individually, the measured input (IP_{1dB}) and output (OP_{1dB}) 1 dB compression points are 23.3 and 15.4 dBm at 24 GHz, and 18.4 and 9.1 dBm at 60 GHz, respectively, as shown in Fig. 5.17a. For concurrent dual-tone input at 24 and 60 GHz, the measured IP_{1dB} and OP_{1dB} are 19 and 11.3 dBm at 24 GHz, and 16.8 and 7.8 dBm at 60 GHz, respectively. The measured P_{1dB}'s at 60 GHz are lower than those at 24 GHz due to higher nonlinear parasitic capacitances of the off-state shunt transistors at 60 GHz. The increased insertion loss occurred beyond the IP_{1dB} point is due to the loss compression resulted from power compression.

Figure 5.17c, d show the measured third-order intercept point (IP_3) for single-band transmitting modes with two tones spaced 100 MHz apart. At 24 GHz, as shown in Fig. 5.17c, the measured IIP_3 and OIP_3 are 31.5 and 24 dBm, respectively. At 60 GHz, the measured IIP_3 and OIP_3 are 31.5 and 24 dBm, respectively, as seen in Fig. 5.17d.

5.3.5 Multi-mode Dual-Band SPDT and T/R Band-Pass Filter-Switches (Design 3)

This section addresses 0.18 μm CMOS dual-band SPDT and T/R band-pass filter-switches, referred to as Design 3, operating in two different frequency bands centered around 24 and 60 GHz. These filter-switches can operate in a variety of separate and concurrent modes, either in single-band, dual-band, transmission, reception, or simultaneous transmission and reception, with band-pass filtering and enhanced isolation. They can also function as a diplexer with switching capability. The T/R filter-switch, especially, allows the transmission and reception in multi-band to be carried out simultaneously with a single antenna—a highly desirable feature for multi-band RF systems, which cannot be achieved with conventional T/R switch that

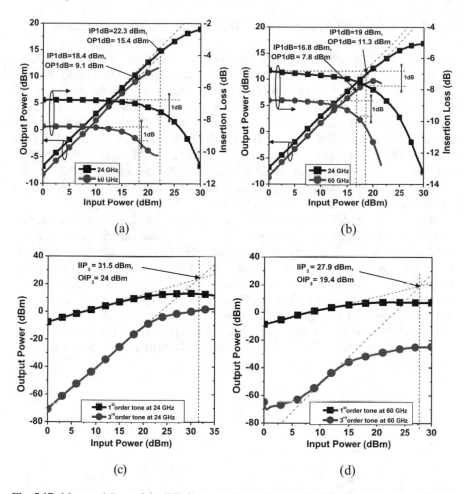

Fig. 5.17 Measured P_{1dB} of the T/R filter-switch with 24 and 60 GHz single-tone input (**a**) and 24/60 GHz concurrent dual-tone input (**b**), and measured IP_3 for single-band mode at 24 GHz (**c**) and 60 GHz (**d**). © [2017] IEEE. Reprinted, with permission, from [5]

can only transmit and receive signals at different times—which could help expand the usage or applications of RF systems. The designs of the 24/60 GHz dual-band band-pass filtering SPDT and T/R switches involve the designs of the individual single-band 24 and 60 GHz band-pass filtering SPST switches. The architectures, designs, and analyses of these single-band band-pass filtering SPST switches and dual-band band-pass filtering SPDT and T/R switches are described in this section. Analytical equations for the switches' insertion losses and isolations are derived and can be used for possible trade-off between the insertion loss and isolation in a design.

5.3.5.1   Single-Band 24 and 60 GHz SPST Band-Pass Filter-Switches

The single-band 24 and 60 GHz SPST band-pass filter-switches are designed based on a second-order band-pass filter with admittance inverters (J-inverters) and parallel resonators as shown in Fig. 5.18a, b, respectively. The band-pass filter design procedure is described in Sect. 2.4. The 24 and 60 GHz filters have 0.01 dB ripple with a fractional bandwidth of 30% centered at 24 and 60 GHz and more than 50 dB rejection at 60 and 24 GHz, respectively. The elements' values for the 24 and 60 GHz filters are $C_{r1,\,r2} = 343$, $165\,f$F and $L_{r1,\,r2} = 120$, $40\,p$H at 24, 60 GHz, respectively, and Z_S, $Z_L = 50$ Ω. The filter would behave as a SPST switch with band-pass filtering characteristics when the capacitors of the filter's resonators are replaced with transistors biased for on- and off-state operations.

Figure 5.18c, d show the schematics of the 24 and 60 GHz band-pass filtering SPST switches, respectively. The series inductor $L^{24}_{j01,\,j23}$ and capacitor $C^{60}_{j01,\,j23}$ in J_{01} and J_{23} of the 24 and 60 GHz band-pass filtering SPST switches have low- and high-pass responses, respectively, prohibiting the 60 and 24 GHz signals from passing through them, respectively. This useful feature, achieved by unique arrangements of the SPST topologies, enables the dual-band band-pass filtering 24/60 GHz SPDT and T/R switches formed by these SPST switches to separate the input signals at 24 and 60 GHz to produce the desired dual-band band-pass filtering responses.

The ABCD matrices of the 24 and 60 GHz SPST switch in Fig. 5.18c, d can be derived as

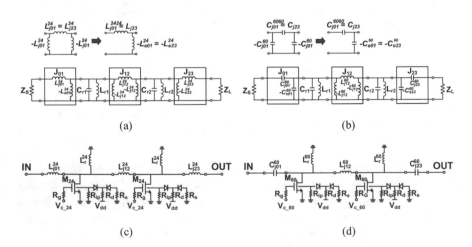

(a)   (b)

(c)   (d)

Fig. 5.18 Schematics of the band-pass filtering SPST switches: **a** 24 GHz band-pass filter and J-inverter implementation, **b** 60 GHz band-pass filter and J-inverter implementation, **c** 24 GHz band-pass filtering SPST switch, and **d** 60 GHz band-pass filtering SPST switch

$$ABCD = \begin{pmatrix} 1 & Z_{j01} \\ 0 & 1 \end{pmatrix} \cdot \begin{pmatrix} 1 & 0 \\ Y_{nMOS} + Y_{L_r} & 1 \end{pmatrix} \cdot \begin{pmatrix} 1 & Z_{j12} \\ 0 & 1 \end{pmatrix}$$

$$\times \begin{pmatrix} 1 & 0 \\ Y_{nMOS} + Y_{L_r} & 1 \end{pmatrix} \cdot \begin{pmatrix} 1 & Z_{j23} \\ 0 & 1 \end{pmatrix} \tag{5.35}$$

where $Z_{j01} = j\omega L_{j01}^{24}$ or $1/j\omega C_{j01}^{60}$, $Z_{j12} = j\omega L_{j12}^{24}$ or $j\omega L_{j12}^{60}$, and $Z_{j23} = j\omega L_{j23}^{24}$ or $1/j\omega C_{j23}^{60}$ are the impedances of J_{01}, J_{12} and J_{23} in the 24 or 60 GHz SPST switch, respectively; $Y_{nMOS} = 1/\left(1/j\omega C_{eq}^{M24, M60} + R_{ch}^{M24, M60}\right)$ is the admittance of shunt transistor M_{24} or M_{60}, respectively, with $R_{ch}^{M24}, C_{eq}^{\cdot M24}$ and $R_{ch}^{M60}, C_{eq}^{M60}$ being the equivalent resistance, capacitance of M_{24} and M_{60}, respectively; and $Y_{L_r} = 1/j\omega L_r^{24,60}$ is the respective admittance of inductors $L_r^{24,60}$.

The transmission coefficient of the 24 and 60 GHz SPST switch can be derived from (5.35) as

$$T = \left| \frac{2Z_0\omega^2 L_r^2 \left(1 + jR_{ch}C_{eq}\omega\right)^2}{\left(A\omega^2 + B\omega + C\right)\left(D\omega^2 + E\omega + F\right)} \right| \tag{5.36}$$

where $A = -C_{eq}L_r\left(R_{ch} + Z_0 + Z_{j01}\right)$, $B = j\left[L_r + R_{ch}C_{eq}\left(Z_0 + Z_{j01}\right)\right]$, $C = Z_0 + Z_{j01}$, $D = -C_{eq}L_r\left[R_{ch}\left(2Z_0 + Z_{j01} + Z_{j12}\right) + Z_{j12}\left(Z_0 + Z_{j01}\right)\right]$, $E = j\left[L_r\left(2Z_0 + 2Z_{j01} + Z_{j12}\right) + R_{ch}C_{eq}L_r Z_{j12}\left(Z_0 + Z_{j01}\right)\right]$, $F = Z_{j12}\left(Z_0 + Z_{j01}\right)$, and Z_0 is the source impedance. Note that there are two different values for C_{eq} (C_{eq}^{M24} or C_{eq}^{M60}), R_{ch} (R_{ch}^{M24} or R_{ch}^{M60}), L_r (L_r^{24} or L_r^{60}), Z_{j01} ($j\omega L_{j01}^{24}$ or $1/j\omega C_{j01}^{60}$), Z_{j12} ($j\omega L_{j12}^{24}$ or $1/j\omega L_{j12}^{60}$) and Z_{j23} ($j\omega L_{j23}^{24}$ or $1/j\omega C_{j23}^{60}$) corresponding to the 24 or 60 GHz SPST switch. T represents the insertion loss (T_{on}) or isolation (T_{off}) of the on- and off-state SPST switches, respectively.

Figure 5.19 shows the insertion losses of the off-state transistors M_{24} ($C_{eq}^{M24} = 340\,fF$) and M_{60} ($C_{eq}^{M60} = 190\,fF$) at 24 and 60 GHz with respect to Q ($Q = 1/\omega R_{ch}^{M24}C_{eq}^{M24}$ and $1/\omega R_{ch}^{M60}C_{eq}^{M60}$), calculated using (5.36) Fig. 5.19 shows that the Q of the off-state transistors employed in the SPST switches affects substantially

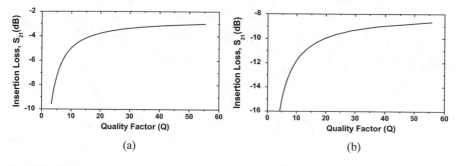

Fig. 5.19 Simulated insertion losses of the 24 and 60 GHz SPST switches versus Q of the off-state transistors: **a** transistor M_{24} at 24 GHz and **b** transistor M_{60} at 60 GHz

the switches' insertion loss and causes a relatively high insertion loss at 60 GHz for M_{60} as compared to that at 24 GHz for M_{24}, where the Q is around 28 as seen in Fig. 5.19b. Equation (5.36) will be used to derive the insertion loss and isolation of the SPDT and T/R band-pass filter-switches to be described in Sects. 5.3.5.2 and 5.3.5.3.

5.3.5.2   24/60 GHz Dual-Band SPDT Band-Pass Filter-Switch

Unlike conventional single-band SPDT switches consisting of two identical SPST switches operating at same frequencies, the 24/60 GHz dual-band SPDT band-pass filter-switch is comprised of two SPST band-pass filter-switches operating at two different frequency bands centered at 24 and 60 GHz as shown in Fig. 5.20. This SPDT filter-switch is part of the T/R filter-switch to be described in Sect. 5.3.5.3. Due to two different pass-bands inherently created by the constituent 24 and 60 GHz SPST band-pass filter-switches, whose topologies are especially configured as described in Sect. 5.3.5.1, the 24/60 GHz dual-band SPDT band-pass filter-switch can also be used as a diplexer with switching function for the 24 and 60 GHz bands. The 24/60 GHz dual-band SPDT band-pass filter-switch can operate in three different modes as follows.

24 GHz Single-Band Operation Mode
The 24 GHz operation mode is described in the equivalent circuits of the 24/60 GHz dual-band SPDT band-pass filter-switch as shown in Fig. 5.21a. It is obtained when transistors M_{24} in the 24 GHz SPST are biased in off-state, which are represented by equivalent capacitors C_{eq}^{M24}, and transistors M_{60} in the 60 GHz SPST are in on-state represented by equivalent channel resistors R_{ch}^{M60}. Under these bias conditions at 24 GHz, the 24 GHz path is approximately matched to the input of the SPDT, while the 60 GHz path approaches an approximate open circuit through the J-inverters J_{01} and J_{23}. Consequently, most of the 24 GHz input signal is transmitted through the 24 GHz SPST switch.

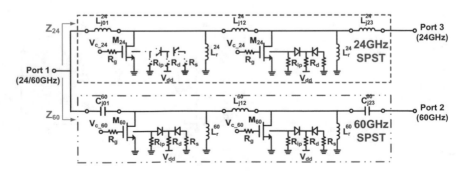

Fig. 5.20   Schematic of the 24/60 GHz dual-band SPDT band-pass filter-switch

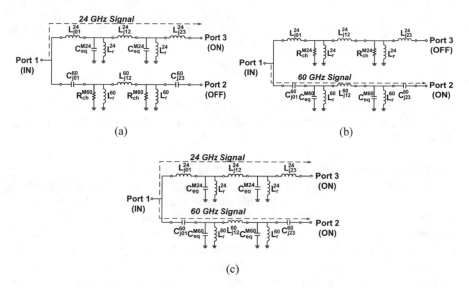

Fig. 5.21 Operations of the 24/60 GHz dual-band SPDT band-pass filter-switch: **a** 24 GHz single-band operation, **b** 60 GHz single-band operation, and **c** 24/60 GHz concurrent dual-band operation

The insertion loss S_{31} and isolation between the two output ports S_{23} can be derived as

$$S_{31} = \frac{Z_{60}^{off}}{Z_{24}^{on} + Z_{60}^{off}} T_{on}^{24} \tag{5.37}$$

$$S_{23} = T_{on}^{24} T_{off}^{60} \tag{5.38}$$

where Z_{24}^{on} and Z_{60}^{off} are the corresponding input impedances Z_{24} and Z_{60}, shown in Fig. 5.20, looking into the on-state 24 GHz path (with M_{24} off) and off-state 60 GHz path (with M_{60} on), respectively, and $T_{on(off)}^{24(60)}$ denotes the insertion loss (T_{on}) or isolation (T_{off}) from (5.36) at 24(60) GHz, respectively.

60 GHz Single-Band Operation Mode

The 60 GHz operation mode can be inferred from Fig. 5.21b corresponding to transistors M_{24} in on-state, represented by equivalent channel resistor R_{ch}^{M24}, and transistors M_{60} in off-state characterized with equivalent capacitor C_{eq}^{M60}. The 60 GHz path is approximately matched to the input of the SPDT at 60 GHz while the 24 GHz path presents an approximate open circuit, hence forcing the majority of the 60 GHz input signal to traverse the 60 GHz SPST switch.

The insertion loss (S_{21}) and isolation between the two output ports (S_{32}) are obtained as

$$S_{21} = \frac{Z_{24}^{off}}{Z_{24}^{off} + Z_{60}^{on}} T_{on}^{60} \tag{5.39}$$

$$S_{32} = T_{off}^{24} T_{on}^{60} \tag{5.40}$$

where Z_{24}^{off} and Z_{60}^{on} are the corresponding input impedances Z_{24} and Z_{60}, as shown in Fig. 5.20, looking into the 24 GHz path with M_{24} on and 60 GHz path with M_{60} off, respectively.

24/60 GHz Concurrent Dual-Band Operation Mode
The 24/60 GHz concurrent dual-band operation mode is described in the equivalent circuit as shown in Fig. 5.21c. In this operation, transistors M_{24} and M_{60} are biased off-state represented by C_{eq}^{M24} and C_{eq}^{M60}. The 24 GHz path simultaneously presents approximate matching and open circuit to the input of the SPDT at 24 and 60 GHz, respectively. On the other hand, the 60 GHz path concurrently provides approximate matching and open circuit to the SPDT's input at 60 and 24 GHz, respectively. As a result, the 24 and 60 GHz signals are routed separately, yet concurrently, through the 24 and 60 GHz SPST switches, respectively.

The insertion losses, S_{31} and S_{21}, and isolation S_{32} can be derived as

$$S_{31} = \frac{Z_{60}^{on}}{Z_{24}^{on} + Z_{60}^{on}} T_{on}^{24} \tag{5.41}$$

$$S_{21} = \frac{Z_{24}^{on}}{Z_{24}^{on} + Z_{60}^{on}} T_{on}^{60} \tag{5.42}$$

$$S_{32} = T_{on}^{24} T_{on}^{60} \tag{5.43}$$

The foregoing qualitative analysis provides the operation principles of the 24/60 GHz dual-band SPDT band-pass filter-switch in three different operating modes and verifies that it can support both concurrent dual-band and single-band operations.

5.3.5.3   24/60 GHz Dual-Band T/R Band-Pass Filter-Switch

Figure 5.22 shows the schematic of the 24/60 GHz dual-band T/R band-pass filter-switch implemented using four SPST filter-switches (or three SPDT filter-switches) [8]. It has five different ports: Port 1 is the transmitting (*TX*) port for both 24 and 60 GHz signals; Port 2 (*ANT1*) and Port 3 (*ANT2*) are the antenna ports for both 24 and 60 GHz; and Port 4 (*RX1*) and Port 5 (*RX2*) are the receiving (*RX*) ports for 24 and 60 GHz signals, respectively. The antenna (*ANT*) ports could belong to a single antenna with two ports operating concurrently at 24 and 60 GHz or two different antennas operating concurrently at 24 and 60 GHz. Table 5.3 shows the 24/60 GHz dual-band T/R band-pass filter-switch's design parameters and their values.

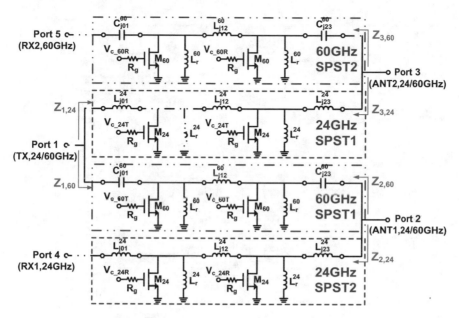

Fig. 5.22 Schematic of the 24/60 GHz dual-band band-pass filtering T/R switch. © [2018] IEEE. Reprinted, with permission, from [8]

Table 5.3 24/60 GHz dual-band T/R band-pass filter-switch's parameters

M_{24}	420 μm/0.18 μm	M_{60}	280 μm/0.18 μm
R_{ch_M24}	1.42 Ω	R_{ch_M60}	2.15 Ω
L_{j01_24}, L_{j23_24}	330 pH	L_{j12_24}	360 pH
C_{j01_60}, C_{j23_60}	60 fF	L_{j12_60}	220 pH
L_{24}	248 pH	L_{60}	35 pH
R_{ip}	10 kΩ	R_d	20 kΩ
R_g	1 kΩ	V_{dd}	1.8 V

The 24/60 GHz dual-band T/R band-pass filter-switch can function in various operating modes consisting of three transmitting modes (24 GHz TX, 60 GHz TX and 24/60 GHz concurrent TX), three receiving modes (24 GHz RX, 60 GHz RX and 24/60 GHz concurrent RX), and four concurrent transmitting and receiving modes (24 GHz TX/RX, 60 GHz TX/RX, 24 GHz TX/60 GHz RX and 24 GHz RX/60 GHz TX). Figure 5.23 shows the concurrent operation modes at 24 GHz (24 GHz TX/RX), 60 GHz (60 GHz TX/RX), and 24/60 GHz (24 GHz RX/60 GHz TX) of the T/R filter-switch and their equivalent circuits. Other operating modes can also be illustrated similarly.

One of the most crucial requirements in multi-band T/R switches designed for concurrent TX and RX operations with a single antenna is the isolation between the TX and RX ports under concurrent operations. Conventional T/R switches, either

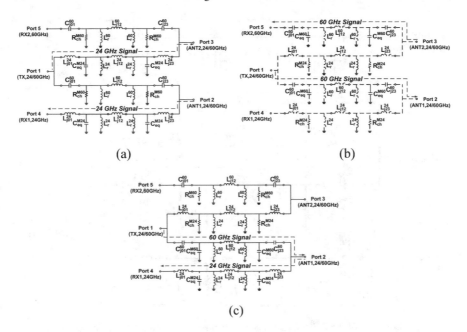

Fig. 5.23 Operations of the 24/60 GHz dual-band band-pass filtering SPDT switch: **a** 24 GHz single-band operation, **b** 60 GHz single-band operation, and **c** 24/60 GHz concurrent dual-band operation. © [2018] IEEE. Reprinted, with permission, from [8]

single- or multi-band operation, are not suitable for concurrent TX and RX operations at the same frequency due the need of turning both the TX and RX paths on at the same time. The proposed 24/60 GHz dual-band T/R band-pass filter-switch overcomes the isolation problem and enables concurrent transmitting and receiving operations due to two reasons. First, different frequency bands are used in adjacent TX and RX ports, hence facilitating increased isolation between TX and RX ports—for example, between the TX port at Port 1 and RX port at Port 5 in Fig. 5.22, in which the 24 GHz SPST1 between Port 1 and Port 3 is operated at 24 GHz, while the 60 GHz SPST2 between Port 3 and Port 5 is operated at 60 GHz. Second, the actual isolation is provided by a combination of the isolation caused by the off-state of the SPST switch and the stop-band rejection of the band-pass filter of the off-state SPST switch. This is particularly useful for CMOS switches at millimeter wave frequencies since increased parasitic capacitances of off-state MOSFETs at these frequencies could limit the isolation [4]. For illustration purpose, we examine the TX/RX concurrent operations at 24 and 60 GHz as shown in Fig. 5.23 in view of the TX-RX isolation.

24 GHz TX/RX Concurrent Operation

The 24 GHz TX/RX concurrent operation mode is described in the equivalent circuit, shown in Fig. 5.23a, which corresponds to the 24 GHz SPST1 and SPST2 and the 60 GHz SPST1 and SPST2 in Fig. 5.22 being on and off, respectively. In this operation, the isolation between Port 1 (*TX* port) and Port 4 (*RX*1 port) is primarily due to the isolation at 24 GHz caused by the 60 GHz SPST1 being in off-state and the rejection at 24 GHz of the band-pass function of the 60 GHz SPST1.

The insertion losses S_{31}, S_{42} and isolation (S_{41}) between Port 1 (*TX* port) and Port 4 (*RX*1 port) can be derived as

$$S_{31} = \left(\frac{Z_{1,60}^{off}}{Z_{1,24}^{on} + Z_{1,60}^{off}} \right) \left(\frac{Z_{3,60}^{off}}{Z_{3,24}^{on} + Z_{3,60}^{off}} \right) T_{on}^{24} \tag{5.44}$$

$$S_{42} = \left(\frac{Z_{2,60}^{off}}{Z_{2,24}^{on} + Z_{2,60}^{off}} \right) T_{on}^{24} \tag{5.45}$$

$$S_{41} = S_{42} S_{21} = S_{42} \left(\frac{Z_{1,24}^{on}}{Z_{1,24}^{on} + Z_{1,60}^{off}} \right) \left(\frac{Z_{2,24}^{on}}{Z_{2,24}^{on} + Z_{2,60}^{off}} \right) T_{off}^{60} \tag{5.46}$$

where $Z_{1(2,3),24(60)}^{on(off)}$ are the impedances denoted in Fig. 5.22 looking into the on (off)-state 24(60 GHz) path at Ports 1, 2 and 3, respectively.

60 GHz TX/RX Concurrent Operation

60 GHz TX/RX concurrent operation mode is described in the equivalent circuit as shown in Fig. 5.23b corresponding to the 60 GHz SPST1 and SPST2 and 24 GHz SPST1 and SPST2 in Fig. 5.22 being on and off, respectively. In this operation, the isolation between Port 1 (*TX* port) and Port 5 (*RX*2 port) is mainly contributed by the off-state of the 24 GHz SPST1 at 60 GHz and the rejection at 60 GHz of the 24 GHz SPST1's band pass filter.

The insertion losses S_{21}, S_{53} and isolation S_{51} between Port 1 (*TX* port) and Port 5 (*RX*2 port) can be derived as

$$S_{21} = \left(\frac{Z_{1,24}^{off}}{Z_{1,24}^{off} + Z_{1,60}^{on}} \right) \left(\frac{Z_{2,24}^{off}}{Z_{2,24}^{off} + Z_{2,60}^{on}} \right) T_{on}^{60} \tag{5.47}$$

$$S_{53} = \left(\frac{Z_{3,24}^{off}}{Z_{3,24}^{off} + Z_{3,60}^{on}} \right) T_{on}^{60} \tag{5.48}$$

$$S_{51} = S_{53} S_{31} = S_{53} \left(\frac{Z_{1,60}^{on}}{Z_{1,24}^{off} + Z_{1,60}^{on}} \right) \left(\frac{Z_{3,60}^{on}}{Z_{3,24}^{off} + Z_{3,60}^{on}} \right) T_{off}^{24} \tag{5.49}$$

24 GHz RX/60 GHz TX Concurrent Dual-Band Operation

The 24 GHz RX/60 GHz TX concurrent dual-band operation mode is described in the equivalent circuit shown in Fig. 5.23c, which corresponds to the 24 GHz SPST2, 60 GHz SPST1 and 24 GHz SPST1, 60 GHz SPST2 in Fig. 5.22 being on and off, respectively. In this operation, the isolation between Port 1 (*TX* port) and Port 4 (*RX*1 port) is mostly determined by the rejections at 24 and 60 GHz of the 60 GHz SPST1 and 24 GHz SPST2's band-pass filters.

The insertion losses S_{21}, S_{42} and isolation S_{41} between Port 1 (*TX* port) and Port 4 (*RX*1 port) can be derived as

$$S_{21} = \left(\frac{Z_{1,24}^{off}}{Z_{1,24}^{off} + Z_{1,60}^{on}} \right) \left(\frac{Z_{2,24}^{on}}{Z_{2,24}^{on} + Z_{2,60}^{on}} \right) T_{on}^{60} \tag{5.50}$$

$$S_{42} = \left(\frac{Z_{2,60}^{on}}{Z_{2,24}^{on} + Z_{2,60}^{on}} \right) T_{on}^{24} \tag{5.51}$$

$$S_{41} = S_{42} S_{21} = S_{42} \left(\frac{Z_{1,24}^{off}}{Z_{1,24}^{off} + Z_{1,60}^{on}} \right) \left(\frac{Z_{2,24}^{on}}{Z_{2,24}^{on} + Z_{2,60}^{on}} \right) T_{on}^{60} \tag{5.52}$$

The 24/60 GHz dual-band SPDT and T/R band-pass filter-switches were designed with 0.18 μm CMOS and fabricated on a TowerJazz 0.18 μm SiGe BiCMOS process [1]. Figure 5.24 shows microphotographs of the SPDT filter-switch and a part of the T/R filter-switch. The fabricated partial T/R filter-switch with 3 ports shown in Fig. 5.24b allows the 60 GHz transmitting and 24 GHz receiving modes to be measured using a 3-port vector network analyzer, and the results can be used to estimate accurately the performance for the 60 GHz receiving and 24 GHz transmitting modes due to the symmetrical structure of the T/R filter-switch.

5.3.5.4 Simulation and Measurement Results

The fabricated 24/60 GHz dual-band SPDT and T/R band-pass filter-switches were measured on-wafer using Rhode and Schwarz vector network analyzer and Cascade probe station.

Figure 5.25 shows the measured and simulated insertion losses, return losses and isolations of the SPDT and T/R switches. The ports corresponding to the measurement parameters are denoted in Fig. 5.24. The results of the SPDT filter-switch are for the 24 and 60 GHz single-band operation modes and the 24/60 GHz concurrent dual-band operation mode as described in Sect. 5.3.5.2.

The results of the T/R filter-switch are for the 24 GHz RX operation, 60 GHz TX operation, and the concurrent dual-band operation of 24 GHz RX (with 2 and 4 as input and output ports, respectively) and 60 GHz TX (with 1 and 2 as input and output ports, respectively) as described in Sect. 5.3.5.3. As mentioned earlier, the results of the concurrent 24 GHz TX and 60 GHz RX operation are similar to those of the 24 GHz RX and 60 GHz TX operation. As can be seen, the results show good agreement between simulations and measurements.

Figure 5.25a, b shows the results for the 24 GHz operation mode. The measured insertion losses of the SPDT (S_{31}^{SPDT}) and T/R (S_{42}^{TR}) switches are 3 and 2.9 dB at 24 GHz, respectively. The measured 3 dB bandwidths of the SPDT and T/R switches are from 14.7–30.3 GHz and 14.6–30.3 GHz, respectively. The measured

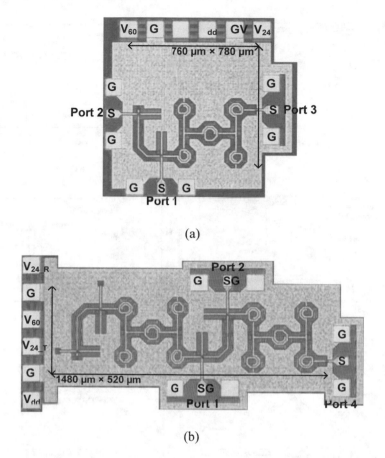

Fig. 5.24 Microphotographs of the 24/60 GHz dual-band band-pass SPDT filter-switch (**a**) and partial T/R filter-switch (**b**). The port numbers in **a**, **b** correspond to those in Fig. 5.22, respectively. © [2018] IEEE. Reprinted, with permission, from [8]

input (S_{11}^{SPDT}, S_{22}^{TR}), output (S_{33}^{SPDT}, S_{44}^{TR}) return losses of the SPDT and T/R filter-switches under on-state are 32, 32 dB and 32, 35 dB at 24 GHz, respectively. The measured isolations between the output ports of the SPDT (S_{32}^{SPDT}) and the TX and RX ports of the T/R (S_{41}^{TR}) switches are 56 and 53 dB at 24 GHz, respectively.

Figure 5.25c, d shows the measured and simulated results for the 60 GHz operation mode. The measured insertion losses of the SPDT (S_{21}^{SPDT}) and T/R (S_{21}^{TR}) switches are 9.4 and 8.7 dB at 60 GHz, respectively. The measured 3 dB bandwidths of the SPDT and T/R switches are 48.5–64.3 GHz and 46.8–62.8 GHz, respectively. The measured input (S_{11}^{SPDT}, S_{11}^{TR}), output (S_{22}^{SPDT}, S_{22}^{TR}) return losses of the SPDT and T/R switches under on-state are 7.5, 12 dB and 7, 11.5 dB at 60 GHz, respectively. The measured isolations between the output ports of the SPDT (S_{32}^{SPDT}) and the TX and RX ports of the T/R (S_{41}^{TR}) filter-switches are 43 dB at 60 GHz.

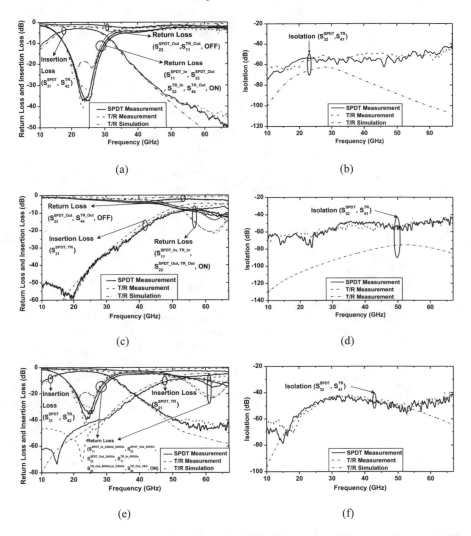

Fig. 5.25 Measured and simulated results of the 24/60 GHz dual-band band-pass filtering SPDT and T/R switches: **a, b** 24 GHz operation, **c, d** 60 GHz operation, and **e, f** 24/60 GHz concurrent operation. © [2018] IEEE. Reprinted, with permission, from [8]

Figure 5.25e, f shows the measured and simulated results for the 24/60 GHz concurrent operation mode. The measured insertion losses of the SPDT ($S_{31}^{SPDT\_24GHz}$, $S_{21}^{SPDT\_60GHz}$) and T/R ($S_{42}^{TR\_24GHz}$, $S_{21}^{TR\_60GHz}$) filter-switches are 3, 9.4 dB and 3, 8.8 dB at 24 and 60 GHz, respectively. It is worth to note that these switches incur no additional loss in the concurrent operating mode as compared to the separate individually operating modes, which validates the concurrent design technique and dictates how well it was executed. The dual-band 3 dB bandwidths of the SPDT and T/R filter-switches are 14.6–30.4 GHz, 48–62.3 GHz and 14.6–30.3 GHz, 48.4–64.8 GHz, respectively. The measured input ($S_{11}^{SPDT\_24GHz\_60GHz}$), output ($S_{33}^{SPDT\_24GHz}$, $S_{22}^{SPDT\_60GHz}$) return losses of the SPDT filter-switch under on-state are 34, 35 dB at 24 GHz and 12, 6.8 dB at 60 GHz, respectively. The measured input ($S_{11}^{TR\_60GHz}$, $S_{22}^{TR\_24GHz}$), output ($S_{22}^{TR\_60GHz}$, $S_{44}^{TR\_24GHz}$) of the T/R filter-switch under on-state are 30, 36 dB at 24 GHz and 11, 7.8 dB at 60 GHz, respectively. From the results ($S_{11}^{SPDT\_24GHz\_60GHz}$ and $S_{22}^{TR\_24GHz\_In}$, $S_{22}^{TR\_60GHz\_Out}$) of the SPDT and T/R filter-switches in Fig. 5.25e, f, it is confirmed that the SPDT and T/R switches have the 24/60 GHz concurrent characteristics. The measured isolations between the output ports of the SPDT (S_{32}^{SPDT}) and the TX and RX ports of the T/R (S_{41}^{TR}) filter-switch are 49, 50 dB at 24 GHz and 50, 57 dB at 60 GHz, respectively. All the results show that the SPDT and T/R filter-switches have similar S-parameter performances as expected.

Figure 5.26a, b show the measured output power and insertion loss versus input power of the SPDT and T/R filter-switches with one tone at 24 and 60 GHz, and two tones at 24 and 60 GHz, respectively. For the single-band 24 GHz operation,

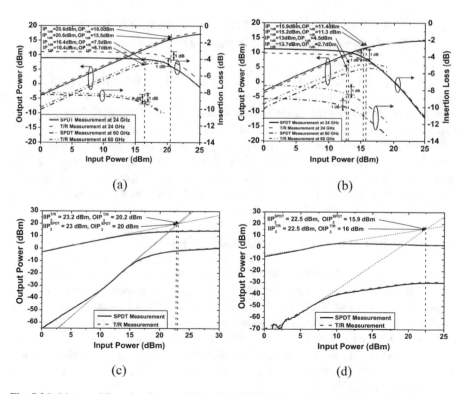

(a)   (b)

(c)   (d)

Fig. 5.26 Measured linearity (P_{1dB} and IP_3) of the 24/60 GHz dual-band SPDT and T/R band-pass filter-switches: **a** measured P_{1dB} at 24 GHz and 60 GHz for single-band operation, **b** measured P_{1dB} at 24 and 60 GHz for concurrent operation, **c** measured IP_3 at 24 GHz, and **d** measured IP_3 at 60 GHz. © [2018] IEEE. Reprinted, with permission, from [8]

the measured input (IP_{1dB}), output (OP_{1dB}) 1 dB power compression points of the SPDT and T/R filter-switches are 20.6, 15.5 dBm and 20.6, 16 dBm at 24 GHz, respectively. For the single-band 60 GHz operation, the measured IP_{1dB}, OP_{1dB} of the SPDT and T/R filter-switches are 16.4, 6.7 dBm and 16.4, 7.5 dBm at 60 GHz, respectively. For the 24/60 GHz concurrent operation, the measured IP_{1dB}, OP_{1dB} of the SPDT and T/R filter-switches are 15.2, 11.3 dBm and 15.9, 11.4 dBm at 24 GHz, and 12.7, 2.7 dBm and 13, 4.5 dBm at 60 GHz, respectively. The reduction in the P_{1dB} at 60 GHz is mainly due to the increased nonlinear parasitic capacitances of the off-state shunt transistors of the switches. Figure 5.26c, d show the measured input IP_3 (IIP_3) and output IP_3 (OIP_3) with two tones spaced 100 MHz apart at 24 and 60 GHz, respectively. At 24 GHz, the measured IIP_3, OIP_3 of the SPDT and T/R filter-switches are 23, 23.2 dBm and 20, 20.2 dBm, respectively. At 60 GHz, they are 22.5, 22.5 dBm and 14.4, 14.5 dBm, respectively. As expected, the SPDT and T/R filter-switches have similar linearity performance from the P_{1dB} and IP_3 measurements.

5.3.6   Wideband Dual-Band SPDT Filter-Switch Utilizing Dual-Band Resonator Concept (Design 4)

This section presents a CMOS concurrent dual-band SPDT filter-switch [9], referred to as Design 4, realized in a 0.18 μm SiGe BiCMOS process that operates concurrently in two different wide bands around 24 and 60 GHz. The SPDT filter-switch is especially configured to operate as a dual-band resonator in the on-state operation for each output path and shows not only switching but also dual band-pass filtering function. The concurrent dual-wideband filter-switch provides decent insertion losses, good power handling, and compact size, even though it operates with integrated band-pass filtering at both frequency bands around 24 and 60 GHz concurrently.

5.3.6.1   Switch Architecture, Design and Analysis

Figure 5.27a shows the schematic of the concurrent dual-wideband SPDT filter-switch. It is realized by two symmetric filter-switching branches, each consisting of series (M_1 or M_3) and shunt (M_2 or M_4) transistors, shunt inductor L_r, and shunt L_1–C_1 or L_2–C_2. Body-floating technique is applied to all the nMOS transistors designed with deep n-well for enhanced isolation and reduced transistors' parasitic capacitances [10, 11]. L_1–C_1 and L_2–C_2 are combined into L_n–C_n connected in shunt at Port 1 between the two switching branches, where $C_n = C_1 + C_2$ and $L_n = L_1/L_2$, as shown in Fig. 5.27b.

Figure 5.27b shows the equivalent circuit of the concurrent dual-wideband SPDT filter-switch when V_c and $\overline{V}_c$ are biased at 1.8 and 0 V, respectively, where R_{on1} and R_{on4} are the on-resistances representing M_1 and M_4 under on-state, respectively,

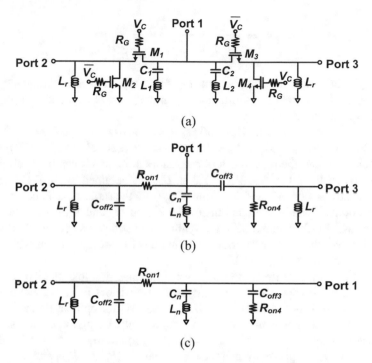

(a)

(b)

(c)

Fig. 5.27 Concurrent dual-wideband SPDT filter-switch: **a** schematic, **b** equivalent circuit when $V_c = 1.8$ V and $\overline{V_c} = 0$ V, and **c** simplified equivalent network between Ports 1 and 2. Reprinted, with permission, from [9]

and C_{off2} and C_{off3} are the off-state capacitances of M_2 and M_3, respectively. In this operation, ports 2 and 3 are on and off, respectively.

Figure 5.27c shows a simplified network between Ports 1 and 2 transformed from the three-port network in Fig. 5.27b. Figure 5.27c is approximately equivalent to a dual-band resonator operating at two distinctive frequencies. To simplify the calculation of the values of the elements, the on-resistances R_{on}'s are neglected, and C_n, L_n, total off-state capacitance $C_{offT} = C_{off2} + C_{off3}$ and L_r can be calculated as [12]

$$C_n = 2\Delta_s/Z_0\omega_s \tag{5.53}$$

$$L_n = 1/\omega_s^2 C_n \tag{5.54}$$

$$
\begin{aligned}
C_{offT} &= C_{off2} + C_{off3} \\
&= \frac{1}{\left[\omega_{c1}^2 + \omega_{c2}^2 - \omega_s^2 - \frac{(\omega_{c1}^2+\omega_{c2}^2)^2-(\omega_{c2}^2-\omega_{c1}^2)^2}{4\omega_s^2}\right] L_n}
\end{aligned} \tag{5.55}
$$

$$L_r = \frac{1}{\left[\omega_{c1}^2 + \omega_{c2}^2 - \omega_s^2 - \frac{1}{L_n C_{offT}}\right] C_{offT}} \qquad (5.56)$$

where Z_0 is the terminating impedance, ω_s is the stop-band center frequency, Δ_s is the stop-band fractional bandwidth, and ω_{c1} and ω_{c2} are the 1st and 2nd pass-band center frequencies, respectively.

From (5.53) to (5.56), with $Z_0 = 50\,\Omega$, $\omega_s = 42$ GHz, $\Delta_s = 0.6$, $\omega_{c1} = 24$ GHz, and $\omega_{c2} = 60$ GHz, $C_n = 92\,f$F, $L_n = 153\,p$H, $C_{offT} = 120\,f$F and $L_r = 170\,p$H. With the calculated C_{offT}, the total width of transistors M_2 and M_3 can be determined as 136 μm. When $V_c = 0$ V and $\overline{V}_c = 1.8$ V, C_{offT} is obtained as $C_{off1} + C_{off4}$ from the off-state transistors M_1 and M_4. Since C_{offT} has the same capacitance in the two bias conditions, it can be seen that the total width of M_2 and M_3 is the same as that of M_1 and M_4. Furthermore, due to the symmetrical structures between Port 1-2 and Port 1-3, the widths of the series transistors M_1 and M_3 are the same as well as those of the shunt transistors M_2 and M_4.

Figure 5.28 presents the trade-off between the insertion loss and isolation at 24 and 60 GHz with respect to the widths of the transistors (M_1, M_3 and M_2, M_4). From Fig. 5.28a, b, the widths of M_1, M_3 and M_2, M_4 for optimal IL and ISO at 24 and 60 GHz are chosen as 64 and 72 μm, respectively. Using these widths, the off-capacitances ($C_{off1} = C_{off3}$ and $C_{off2} = C_{off4}$) are found to be around 55 and 65 fF at 24 and 60 GHz, respectively, and the on-resistances ($R_{on1} = R_{on3}$ and $R_{on2} = R_{on4}$) are around 9.8 and 8.8 Ω at 24 and 60 GHz, respectively. The simulated insertion losses of the filter-switch are 3.2 and 2.8 dB at 24 and 60 GHz, and the simulated isolations of the filter-switch are 30 and 21 dB at 24 and 60 GHz, respectively. All the parameters' values are listed in Table 5.4.

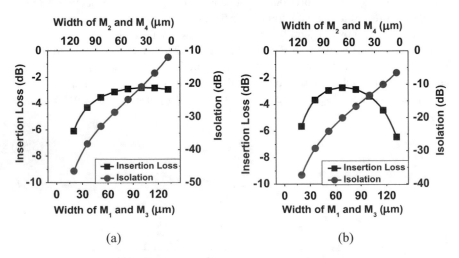

(a)   (b)

Fig. 5.28 Simulated insertion loss and isolation with respect to the width of M_1 and M_3 (or M_2 and M_4) at 24 GHz (**a**) and 60 GHz (**b**). Reprinted, with permission, from [9]

Table 5.4 Dual-wideband SPDT band-pass filter-switch's parameters

M_1, M_3	0.18 μm/64 μm	M_2, M_4	0.18 μm/72 μm
C_n	92 fF	L_n	153 pH
L_r	170 pH	R_G	10 kΩ
C_{off1}, C_{off3}	55 fF	C_{off2}, C_{off4}	65 fF
R_{on1}, R_{on3}	~9.8 Ω	R_{on2}, R_{on4}	~8.8 Ω

Fig. 5.29 Microphotograph of the fabricated concurrent dual-wideband SPDT filter-switch. Port numbers correspond the numbers in Fig. 5.27a. Reprinted, with permission, from [9]

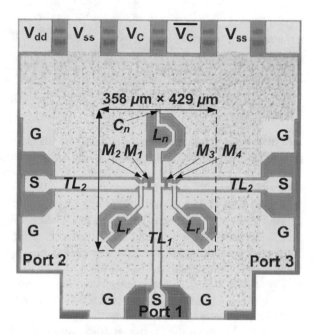

The concurrent dual-wideband SPDT filter-switch was fabricated on TowerJazz 0.18 μm SiGe BiCMOS process [1]. Figure 5.29 shows a microphotograph of the fabricated SPDT filter-switch. The core area occupies 358 μm × 429 μm of the chip space, and the entire chip size including all the dc and RF test pads is 850 μm × 907 μm. In the photograph shown in Fig. 5.29, an extra 50 Ω transmission line (TL$_1$, TL$_2$) is added to each port to avoid possible collision between the RF ports.

5.3.6.2 Simulation and Measurement Results

Figure 5.30 shows the measured and post-layout simulated insertion losses, return losses, and isolations of the concurrent dual-wideband SPDT filter-switch, which show good agreement between them. Due to the symmetrical structure of the switch, the results of the on-state ports 2 and 3 are the same. The measured insertion losses (S_{21}) are 5.4 and 5.2 dB at 24 and 60 GHz, respectively. The measured input (S_{11})

Fig. 5.30 Measured and post-layout simulated results of the concurrent dual-wideband SPDT filter-switch. Reprinted, with permission, from [9]

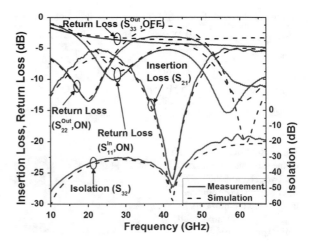

and output (S_{22}) return losses are 9.5 and 8.5 dB at 24 GHz, and 10.6 and 13.2 dB at 60 GHz, respectively. The isolation measured between ports 2 and 3 is 31.4 and 16.5 dB at 24 and 60 GHz, respectively. The peak stop-band rejection is 26 dB at 42.3 GHz.

Figure 5.31 shows the measured output power and insertion loss versus input power of the concurrent dual-wideband SPDT filter-switch. In the case of the 24 GHz single-tone input shown in Fig. 5.31a, the measured input (IP_{1dB}) and output (OP_{1dB}) 1 dB compression points are 20.4 and 13.9 dBm, respectively. For the single tone 60 GHz input, the IP_{1dB} and OP_{1dB} are 17.1 and 10.6 dBm, respectively, as shown in Fig. 5.31b. When the concurrent dual-tone (24/60 GHz) input is injected, the measured IP_{1dB} and OP_{1dB} are 17 and 9.7 dBm at 24 GHz, and 14.5 and 7.4 dBm at 60 GHz, respectively, as seen in Fig. 5.31c.

Figure 5.32 shows the measured third-order intercept points (IP_3) for single-band modes with the two tones spaced 100 MHz apart. At 24 GHz, the measured input IP_3 (IIP_3) and output IP_3 (OIP_3) are 29.4 and 23.6 dBm, respectively, as shown in Fig. 5.32a. At 60 GHz, as seen in Fig. 5.32b, the measured IIP_3 and OIP_3 are 26.8 and 22 dBm, respectively.

5.3.7 Summary of the Developed Dual-Band Filter-Switches

In the foregoing sections, four different filter-switches (Design 1, 2, 3, 4) have been presented. Tables 5.5 and 5.6 summarize the measured performances of these four switches and those of published 24, 40 and 60 GHz single-band [13–20] and 24.5/35 GHz dual-band switches [7].

It is noted that there has been no work reported on 40/60 and 24/60 GHz concurrent dual-band SPDT and T/R switches with dual-band-pass filtering, making it inconclusive in the performance comparisons of the developed dual-band-pass

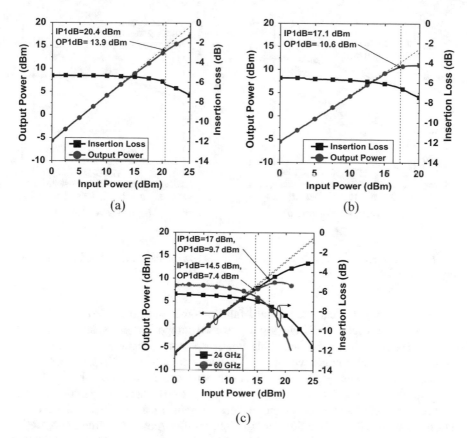

Fig. 5.31 Measured P_{1dB} of the concurrent dual-wideband SPDT filter-switch with 24 GHz single-tone input (**a**), 60 GHz single-tone input (**b**), and 24/60 GHz concurrent dual-tone input (**c**). Reprinted, with permission, from [9]

SPDT and T/R filter-switches. In Design 1, 2 and 3, having 5 ports to accommodate concurrent dual-band and dual-polarization with an integrated dual-band filtering function, the T/R switches have limited design freedom in choosing proper switch's constituent elements, inevitably leading to poorer insertion loss and isolation than those of single-band switches having no filtering and single polarization at 24, 40 or 60 GHz.

It is also noted that the insertion losses at 24, 40 and 60 GHz are high even though the return losses are reasonable. The high insertion loss is due to the two reasons. One is the low Q of the shunt off-state transistors, which were selected to provide the compromise between insertion loss and isolation for both 24/60 and 40/60 GHz simultaneous operations. Another is due to the switch architecture, which provides both concurrent dual-band operation and concurrent switching and filtering functions in a single circuit. The high insertion loss is unfortunately inevitable and

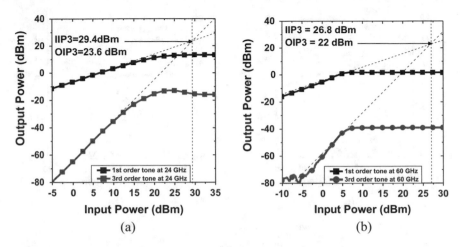

Fig. 5.32 Measured IP$_3$ for single-band modes at 24 GHz (**a**) and 60 GHz (**b**). Reprinted, with permission, from [9]

is the price to pay for designing a multi-function (switching and filtering) and multi-band (concurrent operation in two separate bands) component, especially at such high frequencies as 24, 40 and 60 GHz. This, however, seems acceptable in view of system implementation advantages with multi-function over multi-band for the reduced size and cost using integrated circuits, especially with expensive silicon-based RFICs. Nevertheless, the insertion loss problem at 24, 40 and 60 GHz could be mitigated with the proposed dual-band front-end module described in Fig. 5.33, which consists of the T/R filter-switch, two bidirectional low-noise amplifiers/power amplifiers (LNA/PA), and an antenna. By employing dual-band bidirectional amplifiers similar to that presented in [21], which operate as PA and LNA in the transmitting and receiving modes, respectively, between the T/R switch and the dual-band dual-polarization antenna, enabling high-power and low-noise signals to be transmitted and received in these modes, respectively, the high insertion loss of the T/R switch can be alleviated. The following summarizes and compares the developed filter-switches in Design 1, 2, 3 and 4 in Table 5.5 with the reported switches in Table 5.6.

Design 1 The designed dual-band T/R switch with band-pass filtering has higher insertion loss than the single-band switches without band-pass filtering in [13–16] due to the inclusion of many lumped elements for J-inverters and resonators to embed the dual-band filtering function in the switch as explained earlier. However, the designed T/R filter-switch supports the switching operations at the two distinct 40 and 60 GHz pass-bands concurrently with the stop-band rejection of 30 and 22 dB at 51 GHz for the transmitting and receiving modes, respectively. Also, the isolations at 40 and 60 GHz are all higher than 50 dB both in the transmitting and receiving modes, which are the highest among the reported single-band switches without band-pass filtering operating at 40 or 60 GHz. Moreover, the developed T/R filter-switch is the first reported concurrent 40 and 60 GHz band-pass filtering T/R switch.

Table 5.5 Performance summary of the developed dual-band SPDT and T/R filter-switches

	Design 1		Design 2		Design 3						Design 4	
Process	CMOS on 0.18 μm BiCMOS											
Switching function	TR		TX		SPDT			TR			SPDT/TR	
	TX	RX	TX	X	Single		Dual	Single		Dual	Dual	
Operating Freq. (GHz)	Dual		Dual		24	60	24/60	24	60	24/60	24	60
	40/60	40/60	24/60		24	60	24/60	24	60	24/60	24	60
3 dB BW (GHz)	35.5–44.2/56.4–63.7	35.1–43.7/56.3–63	17.2–27.3/52.5–66.5		14.7–30.3	48.5–64.3	14.6–30.4/48–62.3	14.6–30.3	46.8–62.8	14.6–30.3/48.4–64.8	14.6–30.3/48.4–64.8	–
IL (dB)	10/12.7	8.9/12.5	>6.7/>8.5		3	9.4	3/9.4	2.9	8.7	3/8.8	5.4	5.2
ISO (dB)	57/51	56/51	>18.2/>20.8		56	43	49/50	53	43	50/57	31.4	16.5
ISO-IL (dB)	47/38.3	47.1/38.5	>11.5/>12.3		53	31.6	46/40.6	50.1	32.3	47/48.2	26	11.3
Stop-band Rej. (dB)	33 at 51 GHz	38 at 52 GHz	>40 (36.2–40.8 GHz)	>40 (35.8–41.3 GHz)	23 at 38 GHz			21 at 39 GHz			26 at 42.3 GHz	
OP1dB (dBm)	11.8 (40 GHz input) 1.8 (60 GHz input) 6.4 at 40 GHz, 0 at 60 GHz (40/60 GHz inputs)		15.4 (24 GHz input) 9.1 (60 GHz input) 11.3 at 24 GHz 7.8 at 60 GHz (24/60 GHz inputs)		15.5	6.7	11.4/2.7	16	7.5	11.3/4.5	13.9 (24 GHz input) 10.6 (60 GHz input) 9.7 at 24 GHz, 7.4 at 60 GHz (24/60 GHz inputs)	
IIP3 (dBm)	–		31.5 at 24 GHz 27.9 at 60 GHz		23	22.5	–	23.2	22.5	–	29.4	26.8
Circuit function	Switching and filtering											
Operation mode	TX/RX		TX/RX		**Multiple Single Modes:** 24 GHz TX/24 GHz RX; 60 GHz TX/60 GHz RX **Multiple Concurrent Modes:** 24 GHz TX/RX; 60 GHz TX/RX; 24 GHz TX/60 GHz RX; 24 GHz RX/60 GHz TX; 24 GHz TX/60 GHz TX; 24 GHz RX/60 GHz RX						TX/RX	
Size[a]	1840 μm × 860 μm		1671 μm × 692 μm		760 μm × 780 μm			1480 μm × 520 μm			358 μm × 429 μm	

[a]Estimated size not including dc and RF test pads

Table 5.6 Performance summary of existing single-band 24, 40 and 60 GHz, and dual-band 24.5/35 GHz switches

References	[13]	[14]	[15]	[16]	[17]	[18]	[19]	[20]	[7]
Process	0.13 μm CMOS	0.13 μm CMOS	0.13 μm CMOS	0.13 μm CMOS	0.18 μm CMOS	90 μm CMOS	0.18 μm CMOS	90 μm CMOS	0.18 μm CMOS
Switching function	SPDT/TR	SPDT/TR	SPDT	SPDT/SP4T	SPDT/TR	SPDT/TR	SPDT	SPDT	T/R/C (8 ports)
Freq. (GHz)	40	60	60	60	24	24	60	60	24.5/35
3 dB BW (GHz)	–	–	30–76	50–70	–	–	45–64	–	24–28/31–39
IL (dB)	4.4(TX) 2.7(RX)	5	<2	<2.5/<2.8	6.0	3.4(TX) 3.5(RX)	3.2–3.6	1.6	>9.2/>4.9(TX) >9.4/>9.1(RX)
ISO (dB)	14(TX) 26(RX)	25	>21.1	>30/>20	32.8(RX) 25.5(TX)	22(TX) 16(RX)	>20	>25	>55/>60(TX) >55/>45(RX)
ISO-IL (dB)	9.6(TX) 23.3(RX)	26.4	>19.1	>27.5/>17.2	26.8(RX) 19.5(TX)	18.6 (TX1) 2.5 (RX)	>16.4	>23.4	>45.8/>55.1(TX) >45.6/>35.9(RX)
Stop-band Rej. (dB)	–	–	–	–	–	–	6.5 at 35 GHz	–	>50 at 30 GHz
OP1dB (dBm)	8.4(TX) 7.8(RX)	−1.9	10.8	9.5–10.5	14.5	24.2	>16.4	10.9	1 at 24.5 GHz/4 at 35 GHz
IIP3 (dBm)	–	–	–	–	32.6	–	–	–	23.3 at 24.5 GHz/21.7 at 35 GHz
Circuit function	Switching	Switching	Switching	Switching	Switching	Switching	Filtering Switching	Switching	Filtering Switching

(continued)

Table 5.6 (continued)

References	[13]	[14]	[15]	[16]	[17]	[18]	[19]	[20]	[7]
Operation mode	TX/RX	TX/RX	TX/RX	–	TX/RX	TX/RX	–	–	TX/RX/CAL
Size[a]	800 μm × 500 μm	680 μm × 325 μm	222 μm × 90 μm	390 μm × 320 μm/590 μm × 450 μm	670 μm × 610 μm	–	270 μm × 100 μm	500 μm × 550 μm	–

[a]Estimated size not including the pads

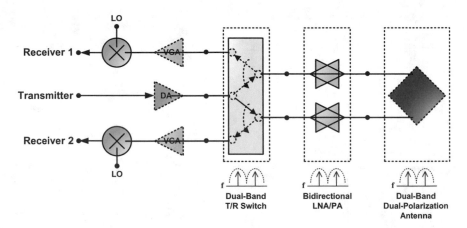

Fig. 5.33 Block diagram of a possible dual-band front-end module

The developed T/R filter-switch has multi-port, with each port handling dual-band signal concurrently, and high isolation between ports, making it attractive for use in dual-band RF systems demanding multi-port and dual-band concurrent operation with band-pass filtering and high isolation. The successful development of this millimetre-wave dual-band band-pass filtering SPDT and T/R switches demonstrates potentials for designing other millimetre-wave multi-pole multi-throw switches capable of filtering functions for silicon-based multi-band RF systems.

Design 2 As for Design 1, the developed T/R filter-switch also has multi-port, with each port being capable of handling dual bands concurrently, making it attractive for use in multi-band RF systems demanding multi-port and multi-band concurrent operation with band-pass filtering function.

Compared to [5] operating at 24.5/35 GHz, the insertion loss of the developed T/R filter-switch is competitive. Compared to the combined insertion losses of a dual-band 24/60 GHz band-pass filter [22] and single-band switches [15–17], the dual-band T/R filter-switch has lower and higher insertion loss at 24 and 60 GHz, respectively. However, this comparison is not conclusive as the T/R filter-switch operates concurrently in dual bands and dual polarizations with integrated filters while [15, 16] operate only in single band and single polarization without filtering. The high insertion loss can be overcome with a front-end module similar to that in [5]. Moreover, the developed T/R filter-switch has very high stop-band rejection (higher than 40 dB) over 36.2–40.8 GHz, as compared to that of [18], and good power handling capabilities.

Design 3 Compared to the switches in [15, 17, 18, 20], which only operate over a single band, either 24 or 60 GHz, and do not have band-pass filtering functions, the dual-band band-pass filtering SPDT and T/R switches can operate in single-band mode at 24 or 60 GHz as well as concurrent dual-band modes at 24 and 60 GHz. The dual-band band-pass filtering T/R switch's operation is more versatile and can be used for various individual and concurrent switching functions. It can operate in single-band transmitting or receiving mode at 24 and 60 GHz and concurrent dual-band transmission or reception at 24 and 60 GHz. It can work completely in concurrent modes including single-band concurrent transmission and reception at 24 or 60 GHz and dual-band concurrent transmission and reception at 24 and 60 GHz. The developed T/R filter-switch makes it possible simultaneous transmission and reception with a single antenna, which is desirable in RF systems yet not feasible with conventional T/R switches. The SPDT and T/R filter-switches can also function as diplexers with switching functions. They also have high isolation between the output ports due to the rejection of the off-state switch as well as the suppression provided by the inherent filtering function. The unique features of the dual-band band-pass filtering SPDT and T/R switches make them attractive for multi-band RF systems requiring single- and multi-band concurrent operations with band-pass filtering.

Design 4 The concurrent dual-wideband SPDT filter-switch has a distinctive dual-band-pass filtering function with good stop-band rejection (26 dB at 42.3 GHz), decent insertion losses at 24 and 60 GHz, and good power handling capability. Moreover, while its size is similar to that of other single-band 60 GHz SPDT switches, the concurrent dual-wideband SPDT switch can also operate in a lower frequency band around 24 GHz.

5.4 Summary

This chapter primarily discusses the design of four different dual-band SPDT and T/R band-pass filter-switches (namely, Design 1, 2, 3 and 4). These switches demonstrate not only switching but also band-pass filtering functions.

Design 1 presents a 0.18 μm CMOS dual-band T/R switch having a band-pass filtering function working concurrently over 35.5–43.7 and 56.4–63 GHz. The developed filter-switch has multiple ports, with each port handling a dual-band signal concurrently, and a high isolation (more than 50 dB over the two pass bands) between ports, which making it attractive for the use of the dual-band RF systems. It is also useful for systems having multi-port and dual-band concurrent operation with band-pass filtering and high isolation. The filter-switch is designed based on a 3rd dual-band band-pass filter with its capacitors replaced with shunt nMOS transistors to accommodate switching function.

Design 2 describes a 0.18 μm CMOS concurrent T/R switch having dual-band-pass filtering function across 17.2–27.3 and 52.5–66.5 GHz. The filter-switch is implemented with dual-band quarter-wavelength LC networks and dual-band resonators with their capacitor replaced with shunt nMOS transistor. The on/off states of the filter-switch are determined through a quarter-wavelength characteristic of the dual-band LC network and the off/on states of the shunt transistor. The developed

T/R filter-switch has multiple ports, with each port being capable of handling dual bands concurrently, making it attractive for the use in multi-band RF systems, and is needed for multi-port and multi-band concurrent operations with band-pass filtering function.

Design 3 shows 0.18 μm CMOS dual-band SPDT and T/R switches having band-pass filtering function over 14.6–30.4 GHz, 48–62.3 GHz and 14.6–30.3 GHz, 48.4–64.8 GHz, respectively. The filter-switches are based on a 2nd order band-pass filter. The dual-band band-pass filtering SPDT can operate in single-band mode at 24 or 60 GHz as well as concurrent dual-band modes at 24 and 60 GHz. The dual-band band-pass filtering T/R switch's operation is more versatile, and it can be used for various individual and concurrent switching functions. It can operate in single-band transmitting or receiving mode at 24 and 60 GHz, and concurrent dual-band transmission or reception at 24 and 60 GHz. It can work completely in concurrent modes including single-band concurrent transmission and reception at 24 or 60 GHz and dual-band concurrent transmission and reception at 24 and 60 GHz. The developed T/R filter-switch makes possible simultaneous transmissions and receptions with a single antenna, which is desirable in RF systems yet not feasible with conventional T/R switches.

Design 4 explains a 0.18 μm CMOS concurrent dual-band SPDT filter-switch operating over two distinctive wide bands around 24 and 60 GHz. In the on-state of the switch, the SPDT filter-switch is equivalent to a dual-band resonator. The concurrent dual-wideband SPDT filter-switch has a decent insertion loss, compact size, and good power handling capability.

References

1. TowerJazz Semiconductor, 4321 Jamboree Road, Newport Beach, CA 92660
2. Um Y, Nguyen C (2017) High-isolation multi-port millimeter-wave CMOS dual-band transmit/receive (T/R) switch with integrated band-pass filtering function. IET Microw Antennas Propag 11(2):253–259
3. Guan X, Ma Z, Cai P, Kobayashi Y, Anada T, Hagiwara G (2006) Synthesis of dual-band bandpass filters using successive frequency transformations and circuit conversions. IEEE Microw Wireless Compon Lett 16(3):110–112
4. Huynh C, Nguyen C (2011) New ultra-high-isolation RF switch architecture and its use for a 10–38-GHz 0.18 μm BiCMOS ultra-wideband switch. IEEE Trans Microw Theory Tech 59(2):345–353
5. Um Y, Nguyen C (2017) A millimeter-wave CMOS dual-bandpass T/R switch with dual-band LC network. IEEE Microw Wireless Compon Lett 27(7):654–656
6. Cheng KM, Wong F (2004) A novel approach to the design and implementation of dual-band compact planar 90° branch-line coupler. IEEE Trans Microw Theory Tech 52(11):2458–2463
7. Lee D, Lee J, Nguyen C (2016) Concurrent dual K/Ka-band T/R/calibration switch module with quasi-elliptic dual-bandpass frequency response implementing metamaterial transmission line and negative resistance. IEEE Trans Microw Theory Tech 64(2):585–598
8. Um Y, Nguyen C (2018) High-isolation multi-mode multi-function 24/60-GHz CMOS dual-bandpass filtering T/R switch. IEEE Microw Wirel Compon Lett 28:696–698
9. Um Y, Nguyen C (2018) Wide-band dual-bandpass 0.18 μm CMOS SPDT switch utilizing dual-band resonator concept. Microw Opt Technol Lett 60:1215–1219

10. Yeh M-C, Tsai Z-M, Liu R-C, Lin KY, Chang Y-T, Wang H (2006) Design and analysis for a miniature CMOS SPDT switch using body floating technique to improve power performance. IEEE Trans Microw Theory Tech 54(1):31–39
11. Bae J, Lee J, Nguyen C (2013) A 10–67-GHz CMOS dual-function switching attenuator with improved flatness and large attenuation range. IEEE Trans Microw Theory Tech 61(12):4118–4129
12. Mao S-G, Wu M-S (2008) Design of artificial lumped-element coplannar waveguide filters with controllable dual-passband responses. IEEE Trans Microw Theory Tech 56(7):1684–1692
13. Yeh MC, Tsai ZM, Liu RC, Lin KY, Chang YT, Wang H (2005) A millimeter-wave wideband SPDT switch with traveling-wave concept using 0.13-m CMOS process. IEEE MTT-S Int Dig 53–56
14. Ta CM, Skafidas E, Evans RJ (2007) A 60-GHz CMOS transmit/ receive switch. IEEE RFIC Symp Dig 725–728
15. He J, Xiong Y Z, Zhang YP (2012) Analysis and design of 60-GHz SPDT switch in 130-nm CMOS. IEEE Trans Microw Theory Tech 60(10):3113–3119
16. Atesal YA, Cetinoneri B, Rebeiz GM (2009) Low-Loss 0.13 μm CMOS 50–70 GHz SPDT and SP4T Switches. In: IEEE radio frequency integrated circuits (RFIC) symposium digital, pp 43–46
17. Ou C-Y, Hsu C-Y, Lin H-R, Chuang H-R (2009) A high-isolation high-linearity 24-GHz CMOS T/R switch in the 0.18 μm CMOS process. In: The European microwave conference, pp 250–253
18. Park P, Shin DH, Yue CP (2009) High-linearity CMOS T/R switch design above 20 GHz using asymmetrical topology and AC-floating bias. IEEE Trans Microw Theory Tech 57(4):948–956
19. Ma K, Mou S, Yeo KS (2013) A miniaturized millimeter-wave standing-wave filtering switch with high P1dB. IEEE Trans Microw Theory Tech 61(4):1505–1514
20. Uzunkol M, Rebeiz GM (2010) A low-loss 50–70 GHz SPDT switches in 90 nm CMOS. IEEE J Solid-State Circ 45(10):2003–2006
21. Kim J, Buckwalter JF (2012) A switchless, Q-band bidirectional transceiver in 0.12 μm SiGe BiCMOS technology. IEEE J Solid-State Circ 47(2):368–380
22. Yeh L-K, Hsu C-Y, Chen C-Y, Chuang H-R (2008) A 24-/60-GHz CMOS on-chip dual-band bandpass filter using trisection dual-behavior resonators. IEEE Electron Device Lett 29(12):1373–1375

Chapter 6
Summary and Conclusions

In the last five chapters, this book has presented the theory, analysis and design of dual-band dual-function CMOS RFIC filter-switches capable of simultaneous switching and filtering for multiband RF systems. Specifically, the developed dual-function filter-switches provide switching in two different band-pass frequency ranges centered at around 40 and 60 GHz and 24 and 60 GHz. The following materials are included in these chapters:

Chapter 1 gives the introduction and background of switches.

Chapter 2 covers the fundamentals of low-pass, high-pass, band-pass, and band-stop filters. It also discusses dual-band band-pass filters employing dual-band resonators, which form the basis for the design of the dual-band band-pass filter-switches presented in this book.

Chapter 3 covers the fundamentals of switches including insertion loss, isolation, power handling and nonlinearity, and figure of merit. Additionally, basic SPST and series, series-shunt SPDT and T/R switches, forming the basis of switch designs, are also addressed.

Chapter 4 presents the fundamentals of switching MOSFETs useful for the design of various switches including general and on/off models as well as operations of MOSFETs. Particular MOSFETs realized using deep n-well for enhanced switching performance are also included. The Q-factor of off-state transistors affecting switch performance, which is often overlooked, is also addressed.

Chapter 5 discusses the design of four different dual-band SPDT and T/R band-pass filter-switches (namely, Design 1, 2, 3 and 4). These switches demonstrate not only switching but also band-pass filtering functions.

Design 1 presents a 0.18-μm CMOS dual-band T/R switch having a band-pass filtering function working concurrently over 35.5–43.7 GHz and 56.4–63 GHz. The developed filter-switch has multiple ports, with each port handling a dual-band signal concurrently, and a high isolation (more than 50 dB over the two pass bands) between ports, which making it attractive for the use of the dual-band RF systems. It is also

C. Nguyen and Y. Um, *Multiband Dual-Function CMOS RFIC Filter-Switches*,
SpringerBriefs in Electrical and Computer Engineering,
https://doi.org/10.1007/978-3-030-46248-2_6

useful for systems having multi-port and dual-band concurrent operation with band-pass filtering and high isolation. The filter-switch is designed based on a 3rd dual-band band-pass filter with its capacitors replaced with shunt nMOS transistors to accommodate switching function.

Design 2 describes a 0.18-μm CMOS concurrent T/R switch having dual-band-pass filtering function across 17.2–27.3 GHz and 52.5–66.5 GHz. The filter-switch is implemented with dual-band quarter-wavelength LC networks and dual-band resonators with their capacitor replaced with shunt nMOS transistor. The on/off states of the filter-switch are determined through a quarter-wavelength characteristic of the dual-band LC network and the off/on states of the shunt transistor. The developed T/R filter-switch has multiple ports, with each port being capable of handling dual bands concurrently, making it attractive for the use in multi-band RF systems, and is needed for multi-port and multi-band concurrent operations with band-pass filtering function.

Design 3 shows 0.18-μm CMOS dual-band SPDT and T/R switches having band-pass filtering function over 14.6–30.4 GHz, 48–62.3 GHz and 14.6–30.3 GHz, 48.4–64.8 GHz, respectively. The filter-switches are based on a 2nd order band-pass filter. The dual-band band-pass filtering SPDT can operate in single-band mode at 24 or 60 GHz as well as concurrent dual-band modes at 24 and 60 GHz. The dual-band band-pass filtering T/R switch's operation is more versatile, and it can be used for various individual and concurrent switching functions. It can operate in single-band transmitting or receiving mode at 24 and 60 GHz, and concurrent dual-band transmission or reception at 24 and 60 GHz. It can work completely in concurrent modes including single-band concurrent transmission and reception at 24 or 60 GHz and dual-band concurrent transmission and reception at 24 and 60 GHz. The developed T/R filter-switch makes possible simultaneous transmissions and receptions with a single antenna, which is desirable in RF systems yet not feasible with conventional T/R switches.

Design 4 explains a 0.18-μm CMOS concurrent dual-band SPDT filter-switch operating over two distinctive wide bands around 24 and 60 GHz. In the on-state of the switch, the SPDT filter-switch is equivalent to a dual-band resonator. The concurrent dual-wideband SPDT filter-switch has a decent insertion loss, compact size, and good power handling capability.

The theory, analysis, and design of dual-band dual-function filter switches presented in this book are useful not only for development of these filter-switches for dual-pass-band at different frequencies, but also for dual-function including switching and another function besides filtering. Moreover, the presented theory, analysis, and design techniques can enable RF engineers to expand further to design multi-band multi-function switches for a variety of specifications and applications as well as for next-generation multi-band RF systems. Successful developments of various dual-band SPDT and T/R filter-switches around 24/60 GHz and 40/60 GHz are evidences that confirm the usefulness and workability of the theory, analysis and design, and possible successes in extending to multi-band and multi-function switches. They indeed pave the way for development of more advanced multi-band multi-function switches and RF systems, especially on CMOS and BiCMOS technologies.

Lastly, we wish to emphasize that, even though this book is relatively concise, it contains sufficient practical and valuable information that should enable RF engineers to successfully design, analysis, and measure multi-band multi-function switches for their own use in many RF applications.

Printed in the United States
By Bookmasters